6th International Conference on Chemistry and Chemical Process (ICCP 2016)

AA004504

MATEC Web of Conferences Volume 49 (2016)

Rome, Italy
1 - 3 February 2016

Editors:

Jacek Namiesnik
Chi-Wai Kan
Dan Yang

ISBN: 978-1-5108-2269-6

Printed from e-media with permission by:

Curran Associates, Inc.
57 Morehouse Lane
Red Hook, NY 12571

Some format issues inherent in the e-media version may also appear in this print version.

Printed by Curran Associates, Inc. (2016)

For additional information, please contact EDP Sciences – Web of Conferences
at the address below.

EDP Sciences – Web of Conferences
17, Avenue du Hoggar
Parc d'Activité de Courtabœuf
BP 112
F-91944 Les Ulis Cedex A
France

Phone: +33 (0) 1 69 18 75 75
Fax: +33 (0) 1 69 28 84 91

contact-edps@webofconferences.org

Additional copies of this publication are available from:

Curran Associates, Inc.
57 Morehouse Lane
Red Hook, NY 12571 USA
Phone: 845-758-0400
Fax: 845-758-2633
Email: curran@proceedings.com
Web: www.proceedings.com

TABLE OF CONTENTS

MASS TRANSFER AND KINETIC STUDIES ON MERCURY ADSORPTION BY TITANIA NANOFIBERS 1
Dixit, Abhilasha; Mishra, P. K.; Alam, M. S.

MAGNETOHYDRODYNAMIC FREE CONVECTION FLOW WITH THERMAL RADIATION AND CHEMICAL REACTION EFFECTS IN THE PRESENCE OF VARIABLE SUCTION 6
Usman, Halima; Uwanta, I. J.; Ahmad, S. K.

HEAVY METAL CONCENTRATIONS OF SELECTED PUBLIC PARKS OF ISTANBUL CITY 12
Demir, Goksel; Ozcan, H. K.; Ozdemir, Huseyin; Pektas, A. O.; Oruc, Ilker; Buyukyildiz, Muhammet

PERFORMANCE OF SUPERCRITICAL METHANOL IN POLYURETHANE DEGRADATION 20
Liu, Lu; Tang, Lihua; Zhu, Zibin; Ni, Yanhui; Wu, Yongqiang

WATER-SENSITIVITY CHARACTERISTICS OF BRIQUETTES MADE FROM HIGH-RANK COAL 26
Geng, Yunguang; Tang, Dazhen; Xu, Hao; Shuling, Tang

MOLECULAR CLONING AND EXPRESSION ANALYSIS OF MEVALONATE PYROPHOSPHATE DECARBOXYLASE IN ANTRODIA CINNAMOMEA 30
Li, Jing; Lin, Xiong-Jie; Shao, En-Si; Lin, Zhan-Xi

CONDUCTIVITY OF CELLULOSE ACETATE MEMBRANES FROM PANDAN DURI LEAVES (PANDANUS TECTORIUS) FOR LI-ION BATTERY 36
Laksono, Endang W.; Marfuatun; Aji, Demas

TRANSPARENT AEROGELS WITH HIGH MECHANICAL STRENGTH COMPOSED OF CELLULOSE-SILICA CROSS-LINKED NETWORKS 40
Zhou, Qing; Shen, Yong; Ai, Sufen; Liu, Bin; Zhao, Yingmin

ESTIMATION OF NO_X EMISSION FROM FOSSIL FUEL CONSUMPTION IN CHINA FOR THE PERIOD 1980-2012 45
Du, Yu; Sun, Xixiang

ENERGETIC EFFICIENCY OF THE VEGETABLE WASTE USED AS SUBSTRATE FOR BIOGAS PRODUCTION 49
Janczak, Damian; Kozlowski, Kamil; Zbytek, Zbyszek; Cieslik, Marta; Bugala, Artur; Czekala, Wojciech

TENSIONED FABRIC STRUCTURES WITH SURFACE IN THE FORM OF CHEN-GACKSTATTER 54
Yee, H. M.; Hadi, M. N. A.

CRUSHED LIMESTONE DUST VALORIZATION FOR ECO FRIENDLY NON FIRED WALL TILES 59
Wangrakdiskul, Ubolrat; Benjaratthapong, Kannika; Jongtrakulsiri, Meatra

NONLINEAR PROGRESSIVE COLLAPSE ANALYSIS INCLUDING DISTRIBUTED PLASTICITY 64
Mohamed, O. A.; Imam, Z. S.; Khattab, Rania

NONLINEAR PROGRESSIVE FAILURE ANALYSIS OF SURROUNDING ROCK SYSTEM BASED ON SIMILARITY THEORY 70
Zhao, Y.; He, L.

DIFFERENTIAL DIFFUSION EFFECTS IN NUMERICAL SIMULATIONS ON SMOKE TOXICITY EVALUATION 74
Pan, Longwei; Jiang, Yong; Wang, Lei; Yin, Lulu; Li, Xu

GREEN SHARING: THE PROPOSED CRITERIA IN GREEN BUILDING STANDARDS TO PROMOTE THE USAGE OF NATURAL HANDICRAFTS IN BUILDING MATERIALS 81
Krasae-In, Aracha

Author Index

Committees

Conference General Chair

Prof. Maciej Baginski,
Faculty of Chemistry, Gdansk University of Technology, Poland

Program Committees Chair

Prof. TANG Zhi-Gang, Tsinghua University, China

Contact Chair

Ms. Flora Feng, Asia-Pacific Chemical, Biological & Environmental Engineering Society

Scientific Editors

Prof. Osman Adiguzel, Firat University, Department of Physics, Turkey

Assoc. Prof. TANG Zhi-Gang, Tsinghua University, China

Assoc. Prof. Sujitra Wongkasemjit, Chulalongkorn University, Thailand

Dr. Claudiu Constantin Manole, University Politehnica Bucharest, Romania

Technical Program Committees

Prof. Nurdan Bahar, Tunceli University, Turkey

Prof. Osman Adiguzel, Firat University, Department of Physics, 23169 Elazig, Turkey

Prof. P.K. Mishra, Indian Institute of Technology (BHU) Varanasi, Varanasi, India

Prof. Goksel Demir, Urban and Regional Planning Dept., Turkey

Dr. M.S. Alam, Department of Chemical Engineering, MNNIT Allahabad, (U.P.), India

Dr. Thitiphan Chimsook, Faculty of Science, Maejo University, Thailand

Prof. Sujitra Wongkasemjit, Chulalongkorn University, Thailand

Dr. C. Constantin Manole, University Politehnica Bucharest, Romania

PREFACE

Welcome to the 2016 International Conference on Chemical Materials and Process (ICCMP 2016) which will be held during May 11-13, 2016, Copenhagen, Denmark. ICCMP 2016 is dedicated to issues related to chemical materials and process.

The major goal and feature of the conference is to bring academic scientists, engineers, industry researchers together to exchange and share their experiences and research results, and discuss the practical challenges encountered and the solutions adopted. Distinguished professors and experts in the area will be invited as Keynote Speakers to deliver the latest information in their respective expertise fields. It will be a golden opportunity for the students, researchers and engineers to interact with the experts and specialists to get their advice or consultation on technical matters, sales and marketing strategies.

This volume presents a selection from papers submitted to the conference from universities, research institutes and industries. All of the papers were subjected to peer-review by conference committee members and international reviewers. The papers selected for this volume depended on their quality and their relevancy to the conference. The volume tends to present to the readers the recent advances in the field of chemical materials and process and various related areas, papers are divided into 7 chapters as follows: electrochemistry, analytical chemistry, analytical chemistry, electrochemistry, electrochemistry, environment & chemistry, civil engineering.

We would like to thank all the authors who have contributed to this volume and also to the organizing committee, reviewers, speakers, chairpersons, sponsors and all the conference participants for their support to ICCMP 2016.

And we're looking forward to seeing you all in ICCMP next year.

Prof. Osman Adiguzel, Firat University, Department of Physics, Turkey
Prof. TANG Zhi-Gang, Tsinghua University, China
Prof. Sujitra Wongkasemjit, Chulalongkorn University, Thailand
Dr. Claudiu Constantin Manole, University Politehnica Bucharest, Romania

MATEC Web of Conferences **64**, 01001 (2016) DOI: 10.1051/matecconf/20166401001
ICCMP 2016

Mass transfer and kinetic studies on mercury adsorption by titania nanofibers

Abhilasha Dixit[1], P. K. Mishra[2] and M. S. Alam[1]

[1]*Department of Chemical Engineering, Motilal Nehru National Institute of Technology, Allahabad, UP, India-211004*
[2]*Department of Chemical Engineering, Indian Institute of Technology (BHU), Varanasi, UP, India-221005*

Abstract. The stringent laws and tight regulation on heavy metal bearing wastewaters being discharged into the water stream lead to emergence of technically advanced and super effective treatment methods. Heavy metal contaminated aqueous streams have become a global concern due to their carcinogenicity and delirious effects on human health. In particular, the health hazards of mercury include adverse effect on central nervous system, pulmonary and kidney functions, etc. Having a maximum discharge limit of 0.001 mg/L, which is the lowest amongst other heavy metals, mercury contamination has become a matter of global concern. In the present study, TiO_2 having a proven track record for its affinity for heavy metals was lab synthesized in the form of fiber mats and tested for its adsorptive behaviour towards Hg(II) in aqueous system. The electrospinned nanofibers possessed smooth morphology and very high surface area $\approx$ 740 m^2/g. The batch adsorption experiments showed titania nanofibers possessed great affinity towards Hg(II). At the most optimized conditions the removal percentage went remarkably high to 95.5% with initial Hg(II) concentration of 0.01 mg/L. Since the rate limiting steps in adsorption are of vital importance in order to define the rate parameters for design purposes, the present study takes into account External mass transfer, Weber and Morris and Boyd mass transfer diffusion models for Hg(II) adsorption on lab synthesized titania nanofibers. The equilibrium data were then analyzed using Langmuir, Freundlich and Temkin sorption models and the characteristic sorption parameters for each isotherm were determined. The Mass transfer mechanism appeared to be film diffusion controlled and data fitted best to Freundlich isotherm with regression value of 0.991. The discussion also focuses on few of the recently used adsorbents for Hg(II) uptake and their comparison with the present study on the basis of removal percentage and kinetics involved. Promising adsorbent characteristics and rapid Hg(II) uptake makes this process a convenient one.

1 Introduction

Water pollution affects the entire biosphere and so is a problem of great concern. The increasing industrialization has played a very important in the degradation of natural water resources. The direct or indirect discharge from the industries into the water bodies demote the water quality and make it unsuitable for aquatic as well human life. Metals reach the environment naturally through the weathering of soils and rocks, and through anthropogenic activities like mining, processing, etc.

Mercury with a maximum discharge limit of 0.001 mg/L is considered to be a global pollutant. Its presence in environment is of significant concern because of its extreme toxicity, persistence and bioaccumulation properties [1]. Once in the environment, mercury gets converted into methyl-mercury chloride, which is its most toxic form. The major industries releasing mercury-laden wastewaters into environment are fertilizers, mining, mineral processing and tanning, etc. human exposure to mercury has severe health affects like neurological imbalance, anxiety, restlessness and lung disease, etc. Techniques like chemical precipitation and ion exchange for heavy metal remediation have their own respective disadvantage of sludge production and being economically infeasible.

Adsorption with its advantage of operation and design flexibility has come up as a very promising technique for heavy metal uptake. Numerous studies have highlighted the efficient use of modified activated carbons [2], chitosan [3], carbon nanotubes [4], etc for heavy metal adsorption. The advantage of using nano adsorbents for heavy metal remediation from wastewaters lies in their properties such as high porosity and enhanced structural properties suitable for adsorption process. Nanostructured metal oxide materials have attracted considerable attention because of their potential applications in many areas such as electronics, photonics, sensors and catalysis [5-7]. Nanosized ceramics with large surface area per unit mass can have enhanced electrical, physical, and chemical properties with respect to their bulk counterparts [8]. Nanofibers are an excellent new class of materials being used for numerous applications such as insulation, medical, cosmetics and filtration, etc. Various approaches for preparation of nano and microstructures of titania have been reported, such as sol-gel processes, pyrolysis, electrospinning, chemical vapor deposition and hydrothermal methods [9-11]. Electrospinning has recently attracted much attention for nano-fiber synthesis since it offers flexibility of control over morphology, porosity, functionality and composition.

The present study takes up the mass transfer and kinetic isothermal model analysis of the Hg^{2+} adsorption results obtained using lab synthesized electrospun titania nanofibers. Mass transfer mechanism is discussed in details via Weber and Morris and Boyd models. The adsorption data is dynamically studied through Langmuir, Freundlich and Temkin isotherm models. A brief comparison with previously published works has been presented.

2 Material and methods

2.1 Nano fiber preparation

Titania nanofibers (TNF) were synthesized using a combination of sol gel and electrospinning techniques. Sol prepared with Titanium Tetraisopropoxide, ethanol and acetone in a fixed ratio was injected to a syringe and the fibers were collected at the collecting plate. After hydrolysis at room temperature, the fibers were calcined in muffle furnace. The fiber synthesis process is described in details in our previous work [12].

2.2 Adsorption experiments

Analytical grade $HgCl_2$ was used to prepare Hg^{2+} stock solution. The Hg^{2+} adsorption studies were carried out in batch mode at varying parameters. Measured quantities of TNF were dosed in Hg^{2+} aqueous solutions, and pH of the solution was varied between 4-8. The aliquots at the outlet were tested for remaining Hg^{2+} at different time intervals using ICP technique. The amount of metal ion at equilibrium, qe and % metal removal is calculated from the mass balance equation given in (1) and (2) respectively [13].

$$qe = \frac{(C_0 - C_e)}{W} \times V \quad (1)$$

$$\% R = \frac{(C_0 - C_e)}{C_0} \times 100 \quad (2)$$

where, V is the volume of the solution, W is the weight of the adsorbent, C_0 and C_e are the initial and equilibrium metal concentrations respectively.

3 Results and discussion

The fiber synthesis carried out with varying process parameters revealed morphological and chemical changes. At the most optimized conditions, perfectly smooth and well-aligned fibers with BET surface area of 740 m^2/g were observed. A surface area as large as this was sufficient enough for heavy metal adsorption application. Fig. 1 represents the smooth surface morphology and average diameter of the synthesized TNF. Fiber morphology and chemical composition results are discussed in details in our previous work [12]. Batch experiments for Hg^{2+} adsorption were performed and effect of various parameters were studied. The pH was varied in the range 4-8, and the maximum removal was obtained at the pH value of 6.

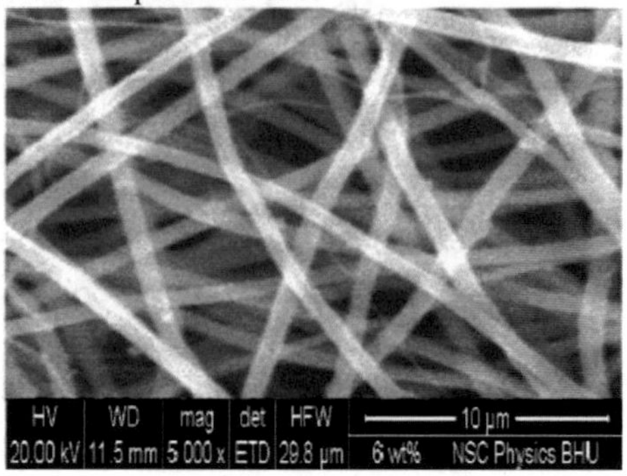

(a)

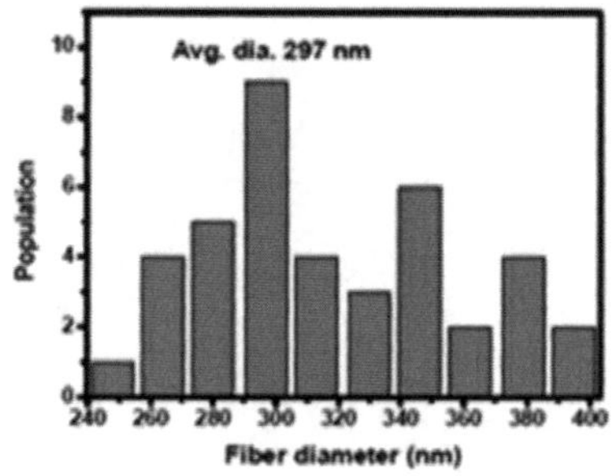

(b)

Figure 1. (a) SEM image (b) average diameter of TNF calcined at 500°C

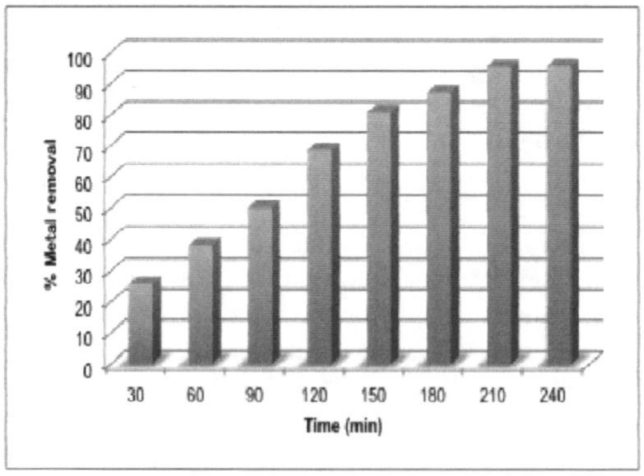

Figure 2. Hg^{2+} adsorption on TNF at different time intervals (TNF 0.05 g/L; pH 6)

3.1 Effect of contact

Fig. 2 shows the change in adsorption efficiency of the system with time, keeping the titania nanofiber dose and system pH to 0.05g/L and 6, respectively. A fast initial

DOI: 10.1051/matecconf/20166401001

adsorption rate is observed up to 150 minutes, which thereafter slows down and remains constant after 210 minutes. This phenomenon is attributed to the initial availability of greater number of vacant sites for metal ions, which apparently decreases as the time increases and finally becomes constant as the equilibrium is achieved.

3.2 Mass transfer mechanism

The rate-limiting step (slowest step of the reaction) is an important factor to be determined so as to establish the mechanism involved for transportation of mass. It may be either the boundary layer (film) or the intra-particle (pore) diffusion of solute on the solid surface from bulk of the solution in a batch process. Different steps viz. solute transfer to the sorbent surface, transfer from the sorbent surface to the active sites and retention on of metal ions on these sites via sorption, complexion and intra-particle precipitation control the sorption kinetics.

Weber and Morris suggested the probability of intra-particle diffusion, as (3). The plot of q_t versus $t^{0.5}$ represents intra-particle diffusion, where q_t represents adsorption capacity at any time t and k_i is intra-particle diffusion rate constant (mg/g min$^{1/2}$).

$$q_t = k_i\sqrt{t} \tag{3}$$

The deviation of the plot from the linearity would indicate the rate-limiting step should be boundary layer (film) diffusion controlled, that is, if the plots are not completely linear, and more-so do not pass through the origin, then intra-particle diffusion could not be the only mechanism involved. In most cases the plot features three stages; initial curved portion due to external mass transfer, followed by an intermediate linear portion due to intra-particle diffusion and a plateau to the equilibrium stage where intra- particle diffusion starts to slow down due to extremely low solute concentrations in the solution. Fig. 3(a) depicts the intra-particle plot for Hg^{2+} adsorption on TNFs. Even though the plot shows an impressive regression value (R^2 = 0.939), it can be seen that the plot is not linear over the whole time range and the graph reflects a dual nature, with initial linear portion followed by plateau. This implies that the external surface adsorption is absent. The stage of intra-particle diffusion is attained and continues till 210 min and finally, equilibrium adsorption starts after 210 min. However, the linear portion of the curve does not pass through the origin and the latter stage of adsorption does not obey Weber–Morris equation. It may be concluded that the adsorption mechanism of these ions from aqueous solution is rather complex process and the intra-particle diffusion was not the only rate-controlling step.

To investigate the mechanism of adsorption, data were fit into Boyd kinetic model [14]. It was originally proposed for intra particle diffusion in spherical particle and is basically a film diffusion model, better known as Boyd's film diffusion model. (4) represents the Boyd kinetic model where F (q_t/q_e) is the fractional attainment of equilibrium at time t, B is the Boyd constant, q_t and q_e are the amounts of metal ion adsorbed at time and at equilibrium, respectively.

$$F = 1 - \frac{6}{\pi^2}\sum_{n=1}^{\infty}\frac{1}{n^2}\,e^{-n^2Bt} \tag{4}$$

Ideally, the mechanism can be called film controlled if the plot of Bt versus time is linear and passes through origin. However, from Fig. 3(b) it can be seen that plot is not completely linear and does not pass through the origin. Also the regression value (R^2 = 0.847) obtained is not very promising. The rate control mechanism of the process does not set well with Boyd model. When a comparison is made between the intra-particle model and Boyd model based on their regression values, intra-particle model shows a better fit. This system seems to follow a complex process with none of the single mechanism taking the complete charge.

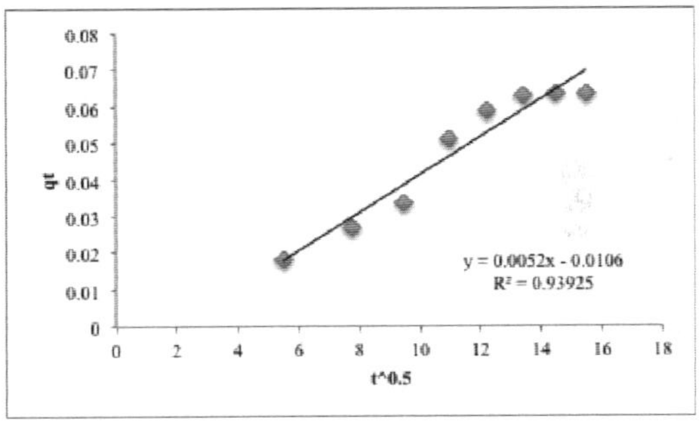

(a)

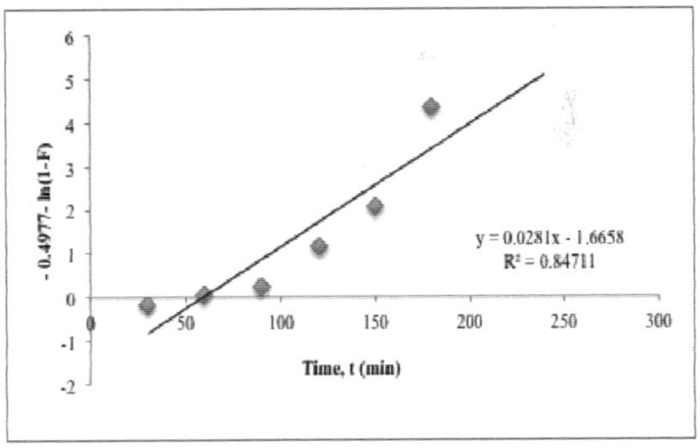

(b)

Figure 3. Mass transfer mechanism model (a) Weber and Morris (b) Boyd

3.3 Adsorption kinetics

Adsorption isotherms are empirical models representing the amount of material bound at the surface as a function of material present in bulk at constant temperature. Langmuir isotherm represents monolayer adsorption and the equation follows Henry's law. It also assumes that each site has same level of energy and the adsorbed molecules have no interaction with each other. Freundlich model assumes multilayer adsorption and Temkin isotherm takes into account adsorbate interactions. Eq (5), (6) and (7) represents Langmuir, Freundlich and Temkin models respectively where, Q_m and Q_e represents the maximum amount of metal ion adsorbed and metal ion adsorbed at equilibrium, C_e is the metal concentration at equilibrium, k_L is Langmuir constant, k_f Freundlich constant and B is Temkin contant.

$$\frac{C_e}{Q_e} = \frac{1}{Q_m k_L} + \frac{C_e}{Q_m} \quad (5)$$

$$Log\ Q_e = Log\ k_f + \frac{1}{n} Log\ C_e \quad (6)$$

$$Q_e = B\ Ln\ A + B\ Ln\ C_e \quad (7)$$

The applicability of these isotherms on adsorption data was adjudged by their regression correlation coefficient, R^2, presented in Table 1. From the table it can be seen that the regression values for both Langmuir and Freundlich are impressive but best linear fit is observed by Freundlich isotherm with highest regression value of 0.991 compared to 0.964 and 0.890 by Langmuir and Temkin models respectively. The factor R_L and 1/n represents the favorability of the process through Langmuir and Freundlich, respectively. When $R_L > 1$, the reaction is unfavorable, R_L = 0-1 represents favorable conditions and R_L = 1 represents linear relationship. The value of favorability factor is close to 1 in case of Langmuir and is well within the favorable limits in case of Freundlich isotherm. The value of constant is also impressive.

Table 1. Isotherm model parameters

Langmuir	
Q_m	6.544
k_L	0.038
R_L	0.989
R^2	0.964
Freundlich	
1/n	0.453
k_f	0.174
R^2	0.991
Temkin	
A	0.054
B	0.241
R^2	0.890

3.4 Comparison with similar works

Numerous earlier works have used nano-sized adsorbents for removing Hg^{2+} from aqueous solutions in batch experiments. Table 2 represents some of the recent works on mercury adsorption in batch mode using nano adsorbents. Manganese chloride nanoparticles [15], functionalized iron nanoparticles [16-18] with particle diameters 100, 40-50 and 10 nm respectively could remove more than 95% Hg(II) at various concentrations. Titania nano particles [19, 20] with diameter ranging from 3-10 nm and large surface area had adsorption capacity of more than 120 mg/g Hg(II) at concentrations ranging from 100-800 mg/L with or without UV illumination. More than 90% of 40-100 ppm Hg(II) in aqueous media was removed by thiol functionalized magnetite nanoparticles [21]. Iron oxide nano composite [22] and nano polymer [23] showed a good affinity over a range of metal ion concentration, with adsorption efficiency around 167.8 and 256.4, respectively. Fe_2O_3-Al_2O_3 nano fibers [24] had a small surface area of 9.6 m^2/g but exhibited Hg(II) adsorption of 89%. Most of the mechanisms were second order reaction and followed Langmuir isotherm. However few works fitted well to Freundlich isotherm as well. The works mentioned above have targeted a higher Hg(II) concentration while the need of the hour is to target the concentrations close to the ones found in real industrial effluents. The lab synthesized TNF used in the present work had very high affinity for Hg^{2+} metal ions. The metal uptake in the present work went to 95.5% at Hg^{2+} concentration of 0.01ppm within 210 minutes. The main reason behind this seems to be the high surface area. Adsorption, being a surface phenomenon largely depends on surface area and with such high surface areas comes higher removal efficiencies.

Table 2. Comparison of recent works on nano adsorbents

Nano Adsorbent	Propert-ies	Working conditio n	Finding s	Ref.
Mn(III)Cl NP on SiO_2/Al_2O_3 mixed oxide support	D = 5-30 SA = 243	C= 0.5-400, A = 0-41.5, pH = 1-12, t = 0-12, T =25-80	Q = 289.5, I-Langmuir Freundlich O-Second	15
Fe_3O_4-Hydroxyapatite NP	D = 100	C = 2.5e^{-8}, A = 0.1-4, pH = 3-9, t = 0-1, T = 10-40, S = 200	R = 98, I-Langmuir O-Second	16
Fe_3O_4 NPs modified with 2-mercaptobenzothiazole	D = 40-50	C = 20-100 ng/L, A = 0.06-2.4, pH = 1-11, t = 0-0.5	R = 98.6 I-Langmuir	17
Polyrhodanine -coated γ-Fe_2O_3 NPs	D = 10 SA = 94.65	C = 1-80, A = 0.5, pH = 2-8, t = 0-12, T = 25	R = 94.5 I-Freundlich O-Second	18
TiO_2 NP	D = 10.19 SA = 98.74	C = 100-800, A = 1-5, pH = 2-9, t = 0.5-9, T = 30-60	Q = 166.66 I-Freundlich, O-Second	19

TiO_2 NPs (+ UV)	D = 9.1 SA > 200	C = 100, A = 10, pH = 6, t = 1.5, T = 25-45	Q = 101 E_a = 25 I-Langmuir O-First	20
TiO_2/montmorillonite (+ UV)	D = 3.1 SA = 239		Q = 123.8 Ea = 22.6 I-Langmuir O-First	
Thiol-functionalized silica-coated magnetite NPs	D = 50	C = 40-100 μg/L, A = 4-8 mg/L, pH = 2-9, t = 0.25, T = 10-35	R = 93.76 I-Freundlich	21
Amino functionalized Fe_3O_4 graphene NC	D = 38.57 SA = 62.43	C = 3, A = 0.16, pH = 1-9, t = 0-5, T = 20-40	Q = 167.8 I-Langmuir O-Second	22
Mercapto functionalized nano-magnetic Fe_3O_4 polymers	D = 40	C = 20-100, A = 0-1, pH = 2-6, t = 0-3, T = 25-45	Q = 256.4, I-Freundlich O-Second	23
Fe_2O_3-Al_2O_3 NF	D = 200-500 SA = 9.6	C = 5-50, A = 2.5, pH = 2-8, t = 0-3, T = RT	R = 89, I-Langmuir O-Second	24

 # NC-Nano composite; NF-Nano fiber; NP-Nano particle; D-Diameter (nm); SA-Surface area (m^2/g) ; C-Hg(II) concentration ; A-Adsorbent dose ; t-time (hours) ; T-Temp (℃) ; Q-Adsorption capacity (mg/g) ;R Removal % ; I-Isotherm ; O-Order

4 Conclusion

This study has shown the applicability of titania nanofibers for Hg^{2+} uptake from aqueous solutions. The adsorption of metal ion was found to be dependent on pH and contact time apart form other factors. The work highlights an enhanced rate of adsorption, which is attributed to the large surface area. The TNFs possessed a surface area of 740 m^2/g characterized by BET. A detailed discussion on mass transfer mechanism, taking into picture both intra particle and film diffusion control is taken up. The nature of the plot indicated that the mass transfer process is rather complex. Langmuir, Freundlich and Temkin isotherms were tested and the constant values, favorability factor and regression values all indicated the suitability of freundlich isotherm model. TNF showed a great affinity for Hg^{2+} and can be a potential candidate for adsorption of other metals as well.

References

1. M. Firouzzare, Q. Wang, Talanta. **101**, 261-266 (2012)
2. K. Kadirvelu, M. Kavipriya, C. Karthika, N. Vennilamani, S. Pattabhi, Carbon. **42**, 745-752 (2004)
3. X. Liu, Q. Hu, Z. Fang, X. Zhang, B. Zhang, Langmuir. **25**, 3-8 (2009)
4. R. O. Long, R. T. Yang, J. Am. Chem. Soc. **123**, 2058-2059 (2001)
5. G. N. Chaudhari, D .R. Bambole, A. B. Bodade, J. Mater. Sci. **41**, 4860-4864 (2006)
6. C. L. Chen, H. S. Weng, Appl. Catal. B-Environ. **55**, 115-122 (2005)
7. S. J. Park, S. Bhargava, E. T. Bender, J. Mater. Res. **23**, 1193-1196 (2008)
8. T. Matsui, M. Harada, Y. Ichihashi, Appl. Catal. A-Gen. **286**, 249-257 (2005)
9. V. Tomer, R. Teye-Mensah, J. C. Tokash, Sol. Energy Mater. Sol. Cells. **85**, 477-488 (2005)
10. S. K. Pradhan, P. J. Reucroft, F. Yang, J. Cryst. Growth. **256**, 83-88 (2003)
11. J. Yang, S. Mei, J. M. F. Ferreira, J. Am. Chem. Soc. **84**, 1696-1702, 2001.
12. A. Dixit, K. Atal, P. K. Mishra, M. S. Alam, Asian J. Chem. **28**, 2, 415-422 (2016)
13. B. H. Hameed, A. M. Din, A. L. Ahmad, J. Hazard. mater. **137**, 3, 695-699 (2006)
14. G. E. Boyd, A. W. Adamson, L. S. Myers-Jr., J. Am. Chem. Soc., **69**, 2836-2848 (1947)
15. M. Arshadi, Chem Eng J. **259**, 170-182 (2015)
16. Kh. Hardani, F. Buazar, K. Ghanemi, M. Kashisaz, M. H. Baghlani-Nezhad, A. Khaledi-Naseb, M. Badri, Amer. Assoc. Sci. Tech. J. Nanosci. **1**, 1, 11-18 (2015)
17. H. Parham, B. Zargar, R. Shiralipour, J. Hazard. mater. **205-206**, 94-100 (2012)
18. J. Song, H. Kong, J. Jang, J. Colloid Interface Sci. **359**, 505-511 (2011)
19. Z. Ghasemi, A. Seif, T. S. Ahmadi, B. Zargar, F. Rashidi, G. M. Rouzbahani, Adv Powder Technol. **23**, 148-156 (2012)
20. B. Dou, V. Dupont, W. Pan, B. Chen, Chem Eng J. **166**, 631–638 (2011)
21. O. Hakami, Y. Zhang, C. J. Banks, Water Res., **46**, 12, 3913-3922 (2012)
22. X. Guo, B. Dua, Q. Wei, J. Yang, L. Hu, L. Yan, W. Xua, J. Hazard. mater. **278**, 211-220 (2014)
23. S. Pan, Y. Zhang, H. Shen, M. Hu, Chem Eng J. **210**, 564-574 (2012)
24. A. Mahapatra, B. G. Mishra, G. Hota, J Hazard Mater. **258- 259**, 116-123 (2013)

 DOI: 10.1051/matecconf/20166401002

Magnetohydrodynamic Free Convection Flow with Thermal Radiation and Chemical Reaction Effects in the Presence of Variable Suction

Halima Usman, Ime Jimmy Uwanta and Sama'ila Kamba Ahmad

Department of Mathematics, Usmanu Danfodiyo University, P.M.B. 2346 Sokoto, Nigeria

Abstract. The aim of the present study is to investigate the effect of flow parameters on the free convection and mass transfer of an unsteady magnetohydrodynamic flow of an electrically conducting, viscous and incompressible fluid past an infinite vertical porous plate in the presence of variable suction. The thermal radiation and chemical reaction effects are assumed to exist within the channel. Non dimensional partial differential equations of governing equations of flow are solved numerically using Crank Nicolson finite difference method. The skin friction, heat and mass transfer rates as well as the effects of various parameters on velocity, temperature and concentration profiles are analyzed. The signifiant results from this study are that an increase in the values of radiation parameter and chemical reaction parameter causes a reduction in the velocity, temperature and concentration.

1 Introduction

The range of free convective flow that occur in nature and in engineering practice is very large and has been extensively considered by many researchers due to its applications in chemically processed industries such as, drying, evaporation at the surface of a water body and food processing. Convection in porous media has applications in geothermal energy recovery, oil extraction, thermal energy storage and flow through filtering devices. Radiative convective flows are encountered in countless industrial and environment processes such as heating and cooling chambers and astrophysical flows. Acharya et al. [1] analyzed free convection and mass transfer flow through a porous medium bounded by vertical infinite surface with constant suction and heat flux. Chamkha [2] studied unsteady MHD convective heat and mass transfer past a semi-infinite vertical moving plate with heat absorption. Cosiderable research work has been published on the study of radiative convective flows in a variety of geometrical configurations with numerical and mathematical models. Chandrakala and Antony [3] discussed radiation effects on MHD flow past an impulsively started infinite vertical plate with variable temperature. Duwairi and Damesh [4] presented a study on heat transfer from radiative vertical porous surfaces. Unsteady free convection flow past a vertical porous plate was investigated by Helmy [5]. Hossain and Takhar [6] reported radiation effect on mixed convection along a vertical plate with uniform surface temperature.

In recent years, progress has been considerably made in the study of heat and mass transfer with chemical reaction due to its applications in many branches of science and engineering. Kandasamy et al. [7] analyzed simultaneously the effects of chemical reaction, heat and mass transfer along a wedge with heat source and concentration in the presence of suction or injection. Kawala and Odda [8] obtained numerical results for the effect of chemical reaction on unsteady natural convection boundary layer flow over a semi-infinite vertical cylinder. Recently, a detailed numerical study was reported by Kishore et al. [9] on the effects of chemical reaction, radiation and magnetic field on the unsteady free convection flow, heat and mass transfer characteristics in a viscous, incompressible and electrically conducting fluid past an exponentially accelerated vertical plate. Mahdy [10] performed an analysis to investigate the effects of chemical reaction and heat generation or absorption on double diffusive convection from vertical truncated cone in a porous media with variable viscosity. Mohammed et al. [11] investigated the effect of similarity solution for MHD flow through vertical porous plate with suction. Patil and Chamkha [12] discussed heat and mass transfer from mixed convection flow of polar fluid along a plate in porous media with chemical reaction. In another article, Rajeshwari *et al.* [13] presented the effect of chemical reaction on the forced and free convection boundary layer that flows in a semi-infinite expanse of an electrically conducting viscous incompressible fluid past a vertical porous plate.

Rajput and Kumar [14] presented the effect of radiation on MHD flow past an impulsively started vertical plate with variable heat and mass transfer. Raptis and Perdikis [15] described radiation and free convection flow past a moving plate. Saleh et al. [16] examined convective heat and mass transfer characteristics of an incompressible MHD visco-elastic fluid flow immersed in a porous medium over a stretching sheet with chemical reaction and thermal stratification. Sarada and Shankar [17] presented a study of chemical reaction effect on an unsteady MHD convective flow past an infinite vertical moving plate embedded in a porous medium with heat source.

Sattar [18] took an account on the effect of free and forced convection boundary layer flow through a porous medium with large suction. Soundalgekar and Takher [19] discussed the effect of radiation on the natural convection flow of a gas past a semi-infinite plate using the Coghy-Vincentine-Gilles equilibrium model. Rahman [20] discussed the combined effects of internal heat generation and higher order chemical reaction on the non-Darcian forced convective flow of a viscous incompressible fluid with variable viscosity and thermal conductivity over a stretching surface embedded in a porous medium.

The objective of this paper is to study the effects of thermal radiation and chemical reaction on an unsteady MHD free convection heat and mass transfer flow past an infinite vertical plate in presence of variable suction. In this study, the effects of different flow parameters encountered in the equations are also studied. The problem is solved numerically using the implicit finite difference method of Crank Nicolson type, which is more economical from the computational point of view.

2 Mathematical analysis

An unsteady MHD free convection heat and mass transfer flow of an electrically conducting, viscous incompressible and radiating fluid over a porous vertical infinite plate with chemical reaction and thermal conductivity is considered. In the Cartesian coordinate, the x' - axis is assumed to be along the infinite plate in the direction of the flow and the y' - axis is taken normal to it. A magnetic field B_0 of uniform strength is applied transversely to the direction of the flow as shown in Fig. 1. Initially, the plate and the fluid are at the same temperature T_∞ in a stationary condition with concentration level C_∞ at all points. For $t' > 0$, the temperature and concentration at the plate $y' = 0$ is raised to T'_w and C'_w respectively and that of $y' \to \infty$ is lowered to T'_∞ and C'_∞ such that $T'_w > T'_\infty$ and $C'_w > C'_\infty$. It is assumed that the induced magnetic field is negligible in comparison with the applied one which corresponds to a very small magnetic Reynolds number. The flow has significant thermal radiation, chemical reaction and suction effects.

Within the above framework, the governing boundary layer equations of continuity, momentum, energy and concentration under Boussinesq's approximation are as follows:

$$\frac{\partial v'}{\partial y'} = 0 \tag{1}$$

$$\frac{\partial u'}{\partial t'} - v_0\frac{\partial u'}{\partial y'} = \nu\frac{\partial^2 u'}{\partial y'^2} - \left(\frac{\sigma B_0^2}{\rho} + \frac{\nu}{k^*}\right)u' + g\beta(T' - T'_\infty) + g\beta^*\left(C' - C'_\infty\right) \tag{2}$$

$$\frac{\partial T'}{\partial t'} - v_0\frac{\partial T'}{\partial y'} = \frac{k_0}{\rho C_p}\frac{\partial}{\partial y'}\left[1 + \alpha(T' - T'_\infty)\frac{\partial T'}{\partial y'}\right] - \frac{1}{\rho C_p}\frac{\partial q_r}{\partial y'} \tag{3}$$

$$\frac{\partial C'}{\partial t'} - v_0\frac{\partial C'}{\partial y'} = D\frac{\partial^2 C'}{\partial y'^2} - R^*(C' - C'_\infty)^n \tag{4}$$

The initial and boundary conditions are

$$\begin{aligned} &t \le 0: u' = 0,\ T' \to T'_\infty,\ C' \to C'_\infty \ for\ all\ y' \\ &t > 0: u' = 0,\ T' \to T'_w,\ C' \to C'_w \ at\ y' = 0 \\ &\qquad u' = 0,\ T' \to T'_\infty,\ C' \to C'_\infty \ as\ y' \to \infty \end{aligned} \tag{5}$$

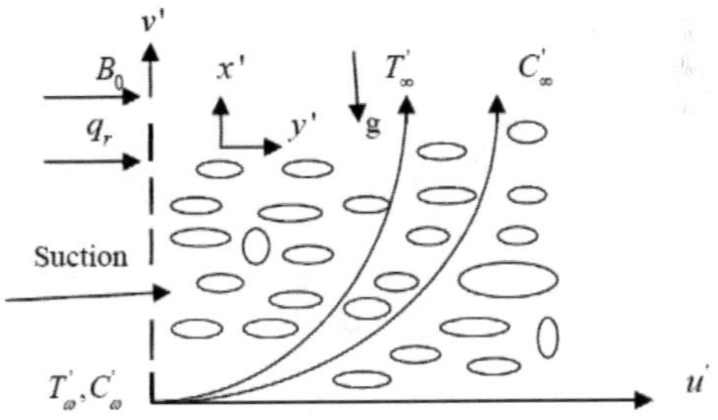

Figure 1. Physical configuration of the problem.

From the continuity equation, it can be seen that v' is either a constant or a function of time. Hence, on integrating equation (1), the suction velocity normal to the plate is assumed in the form:

$v' = -v_0$, where v_0 is the mean suction velocity which is non-zero positive constant. The negative sign indicates that the suction velocity is directed towards the plate, ν is the kinematic viscosity, g is the gravitational force, β is the coefficient of thermal expansion, ρ the density of the fluid, σ is the electrical conductivity of the fluid, β^* is the coefficient of concentration expansion, C_p is the specific heat capacity at constant pressure, k^* is the permeability of the porous medium, k_0 is the thermal conductivity of the fluid, α is a constant depending on the nature of the fluid, R^* is the dimensionless chemical reaction, n is the order of chemical reaction, D is the coefficient of mass diffusivity, q_r is the radiative heat flux in the y - direction, B_0 is the electromagnetic induction.

Using the following non-dimensional quantities:

$$
\begin{aligned}
&u=\frac{u'}{u_0},\ t=\frac{t u_0}{\nu},\ y=\frac{u_0 y'}{\nu},\ \theta=\frac{T'-T'_\infty}{T'_w-T'_\infty},\\
&C=\frac{C'-C'_\infty}{C'_w-C'_\infty},\Pr=\frac{\nu\rho C_p}{k_0},Sc=\frac{\nu}{D},k=\frac{k^* u_0^2}{\nu^2},\\
&M=\frac{\sigma B_0^2 \nu}{\rho u_0^2},Gr=\frac{\nu g\beta\left(T'_w-T'_\infty\right)}{u_0^3},\\
&Gc=\frac{\nu g\beta^*\left(C'_w-C'_\infty\right)}{u_0^3},\lambda=\alpha\left(T'-T'_\infty\right),\gamma=\frac{\nu_0}{u_0},\\
&K_r=\frac{R^*\nu}{u_0^2},\ R=\frac{16 a\sigma'_0 \nu T'^3_\infty}{k' u_0^2}.
\end{aligned}
\tag{6}
$$

Using (6), the equations (1) - (4) take the following form:

$$
\begin{aligned}
\frac{\partial u}{\partial t}-\gamma\frac{\partial u}{\partial y}=\frac{\partial^2 u}{\partial y^2}-\left(M+\frac{1}{k}\right)u\\
+Gr\theta+GcC
\end{aligned}
\tag{7}
$$

$$
\begin{aligned}
\frac{\partial \theta}{\partial t}-\gamma\frac{\partial \theta}{\partial y}=\frac{\lambda}{\Pr}\left(\frac{\partial \theta}{\partial y}\right)^2\\
+\frac{1}{\Pr}(1+\lambda\theta)\frac{\partial^2\theta}{\partial y^2}-\frac{R}{\Pr}\theta
\end{aligned}
\tag{8}
$$

$$
\frac{\partial C}{\partial t}-\gamma\frac{\partial C}{\partial y}=\frac{1}{Sc}\frac{\partial^2 C}{\partial y^2}-K_r C^n
\tag{9}
$$

The initial and boundary conditions in dimensionless form are:

$$
\begin{aligned}
&t\le 0:\ u=0,\ \theta=0,\ C=0,\ for\ all\ y\\
&t>0: u=0,\ \theta=1,\ C=1,\ at\ y=0\\
&\qquad u=0,\ \theta=0,\ C=0\ as\ y\to\infty
\end{aligned}
\tag{10}
$$

where $\Pr, Sc, M, Gr, Gc, \lambda, \gamma, R$ and K_r are Prandtl number, Schmidt number, Magnetic field parameter, thermal Grashof number, mass Grashof number, variable thermal conductivity, variable suction parameter, radiation parameter and chemical reaction parameter respectively. Of practical interest in many applications of the problem under consideration are the skin friction coefficient, the heat transfer coefficient and the mass transfer coefficient, which are given as

$$
C_f=\left(\frac{\partial u}{\partial y}\right)_{y=0},\quad Nu=-\left(\frac{\partial \theta}{\partial y}\right)_{y=0},\quad Sh=\left(\frac{\partial C}{\partial y}\right)_{y=0}
$$

3 Numerical solutions

The system of transformed coupled non-linear partial differential equations (7), (8) and (9) together with the initial and boundary conditions (10) have been solved numerically using a finite difference technique of implicit type namely Crank-Nicolson implicit finite difference method. The equivalent finite difference approximations corresponding to equations (7)-(9) are given as follows:

$$
\begin{aligned}
&\left(\frac{u_{i,j+1}-u_{i,j}}{\Delta t}\right)-\frac{\gamma}{2\Delta y}\left(u_{i+1,j}-u_{i-1,j}\right)=\\
&\frac{1}{2\left(\Delta y\right)^2}\left(u_{i+1,j+1}-2u_{i,j+1}+u_{i-1,j+1}+u_{i+1,j}-2u_{i,j}+u_{i-1,j}\right)\\
&-\left(M+\frac{1}{k}\right)u_{i,j}+Gr\left(\theta_{i,j}\right)+Gc\left(C_{i,j}\right)
\end{aligned}
\tag{11}
$$

$$
\begin{aligned}
&\left(\frac{\theta_{i,j+1}-\theta_{i,j}}{\Delta t}\right)-\frac{\gamma}{2\Delta y}\left(\theta_{i+1,j}-\theta_{i-1,j}\right)=\\
&\frac{H}{2\Pr(\Delta y)^2}\left(\theta_{i+1,j+1}-2\theta_{i,j+1}+\theta_{i-1,j+1}+\theta_{i+1,j}-2\theta_{i,j}+\theta_{i-1,j}\right)\\
&+\frac{\lambda}{\Pr(\Delta y)^2}\left(\theta_{i+1,j}-\theta_{i,j}\right)^2-\frac{R}{\Pr}\left(\theta_{i,j}\right)
\end{aligned}
\tag{12}
$$

$$
\begin{aligned}
&\left(\frac{C_{i,j+1}-C_{i,j}}{\Delta t}\right)-\frac{\gamma}{2\Delta y}\left(C_{i+1,j}-C_{i-1,j}\right)=\\
&\frac{1}{2Sc(\Delta y)^2}\left(C_{i+1,j+1}-2C_{i,j+1}+C_{i-1,j+1}+C_{i+1,j}-2C_{i,j}+C_{i-1,j}\right)\\
&-K_r\left(C_{i,j}\right)^n
\end{aligned}
\tag{13}
$$

the initial and boundary conditions are :

$$
\begin{aligned}
&u_{i,j}=0,\ \theta_{i,j}=0,\ C_{i,j}=0\ \ for\ all\ j=0\\
&u_{0,j}=0,\ \theta_{0,j}=1,\ C_{0,j}=1\\
&u_{h,j}=0,\ \theta_{h,j}=0,\ C_{h,j}=0
\end{aligned}
\tag{14}
$$

Equations (11), (12) and (13) are simplified as follows

$$
\begin{aligned}
-U_1 u_{i-1,j+1}+U_2 u_{i,j+1}-U_1 u_{i+1,j+1}=U_3 u_{i-1,j}\\
+U_4 u_{i,j}+U_5 u_{i+1,j}+U_6\theta_{i,j}+U_7 C_{i,j}
\end{aligned}
\tag{15}
$$

$$
\begin{aligned}
-T_1\theta_{i-1,j+1}+T_2\theta_{i,j+1}-T_1\theta_{i+1,j+1}=T_3\theta_{i-1,j}\\
+(T_4-T_7)\theta_{i,j}+T_5\theta_{i+1,j}+T_6(\theta_{i+1,j}-\theta_{i,j})^2
\end{aligned}
\tag{16}
$$

$$
\begin{aligned}
-V_1 C_{i-1,j+1}+V_2 C_{i,j+1}-V_1 C_{i+1,j+1}=V_3 C_{i-1,j}\\
+V_4 C_{i,j}+V_5 C_{i+1,j}-V_6(C_{i,j})^n
\end{aligned}
\tag{17}
$$

where

$r_1=\frac{\gamma\Delta t}{\Delta y}, r_2=\frac{\Delta t}{(\Delta y)^2}, r_3=2\Delta t, r_4=M+\frac{1}{k}, H=1+\lambda\theta, U_1=r_2,$

$U_2=2(1+r_2), U_3=r_2-r_1, U_4=2-2r_2-r_3 r_4, U_5=r_1+r_2,$

$U_6=r_3 Gr, U_7=r_3 Gc, V_1=r_2, V_2=2(Sc+r_2), V_3=r_2-r_1 Sc,$

$V_4=2(Sc-r_2), V_5=r_1 Sc+r_2, V_6=r_3 ScKr, T_1=Hr_2,$

$T_2=2(\Pr+Hr_2), T_3=Hr_2-r_1\Pr, T_4=2(\Pr-Hr_2), T_5=r_1\Pr+Hr_2,$

$T_6=2\lambda r_2, T_7=r_3 R.$

DOI: 10.1051/matecconf/20166401002

4 Results and discussion

In order to get a physical view of the present problem, numerical results for velocity profiles (u), temperature profiles (T), concentration profiles (C), skin-friction (C_f), rate of heat transfer in terms of Nusselt number (Nu) and rate of mass transfer in terms of Sherwood number (Sh) have been computed with help of graphs by assigning numerical values for different values of pertinent parameters. The following values are used for the investigation: $Gr = Gc = 2.0, M = K_r = k = t = 0.1,$

$$\Pr = 0.71, Sc = 0.6, R = 3.0, n = 5.0 \text{ and } \lambda = \gamma = 0.5$$

Fig. 2 and 3 depicts the fluid velocity distribution profiles with variation of porosity parameter (k) and magnetic parameter (M). On increasing k, it is analyzed that the fluid velocity distribution in this figure increases. It can be seen in Fig. 3 that increasing the magnetic parameter (M) causes a fall in the velocity of the fluid. The effect of transverse magnetic field on an electrically conducting fluid slows down the motion of the fluid. Fig. 4, 5, 6, 7 and 8 plot the variation of the fluid velocity, temperature and concentration distributions for various values of suction parameter (γ) and chemical reaction parameter (K_r). It is observed that as γ and K_r increases, both fluid velocity, temperature distribution and concentration distribution decreases respectively. Fig. 9, 10 and 11 presented the growth of the fluid boundary layer and the thermal distribution profiles with the variation of radiation parameter (R) and Prandtl number $(\Pr)$ respectively. It is observed that increasing either R or Pr produces a significant decrease in the velocity and thermal condition of the fluid. Fig. 12 shows that increasing Schmidt number (Sc) tends to decrease the fluid concentration boundary layer thickness. Fig. 13, 14 and 15 represent the variation of skin friction against magnetic parameter (M), the Nusselt number against radiation parameter (R) and the Sherwood number against chemical reaction parameter (K_r) for various values of Prandtl number ($\Pr$). It is evident from Fig. 13 that higher values of M and Pr reduce the skin friction while the Nusselt number tend to decrease rapidly as the Prandtl number is increased and the Sherwood number increases with increasing Prandtl number.

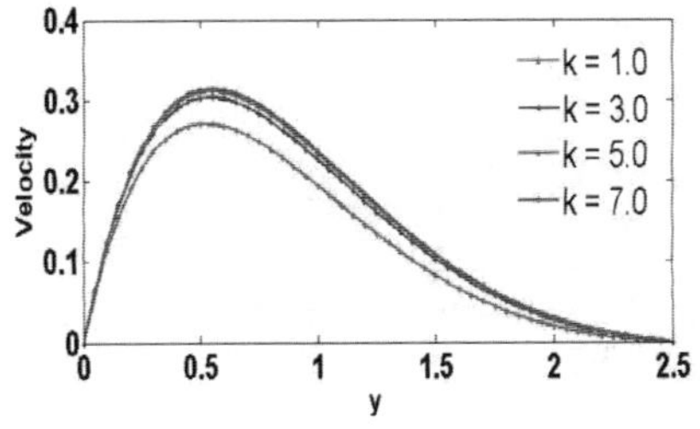

Figure 2. Variation of velocity with (k).

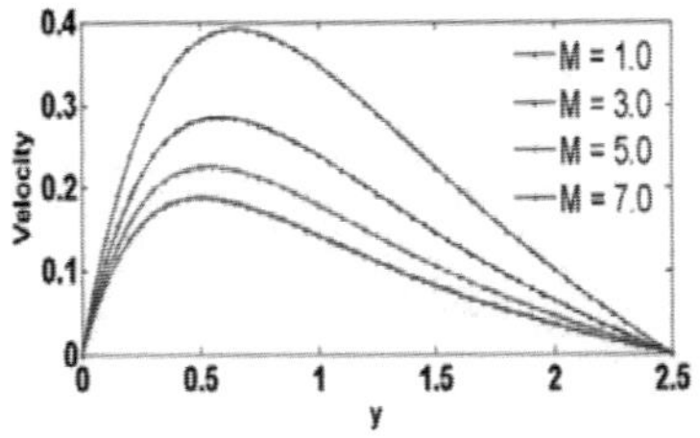

Figure 3. Variation of velocity with (M).

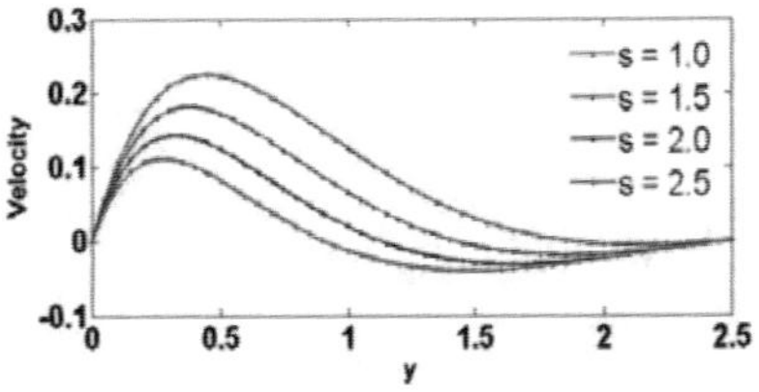

Figure 4. Variation of velocity with ($\gamma = s$).

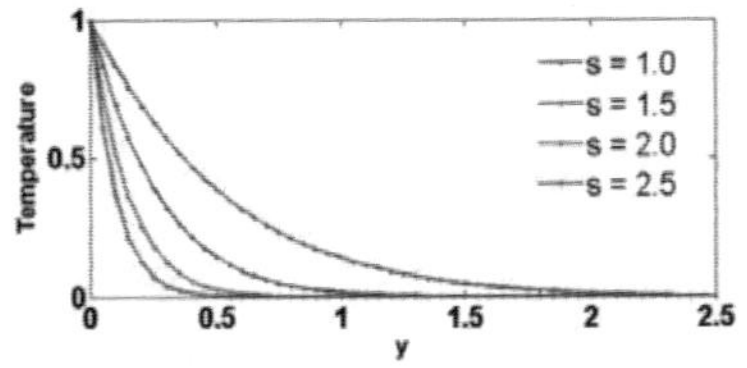

Figure 5. Variation of temperature with ($\gamma = s$).

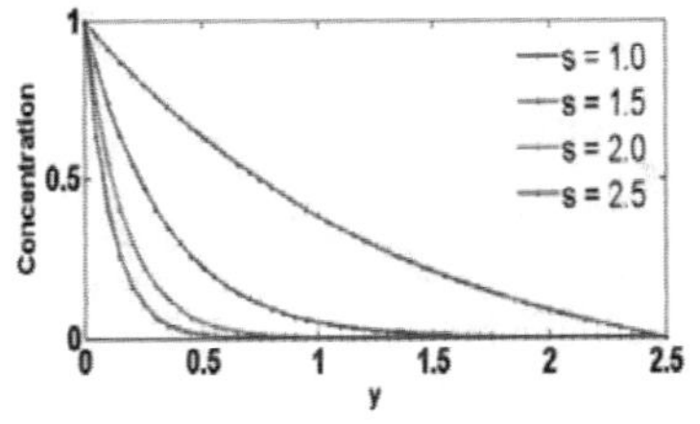

Figure 6. Variation of concentration with ($\gamma = s$).

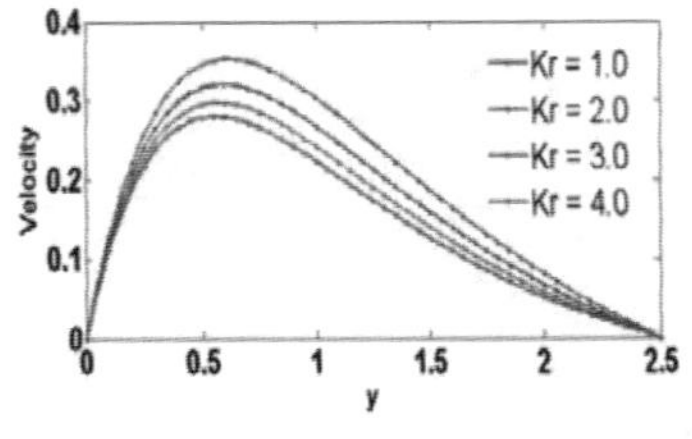

Figure 7. Variation of velocity with (K_r).

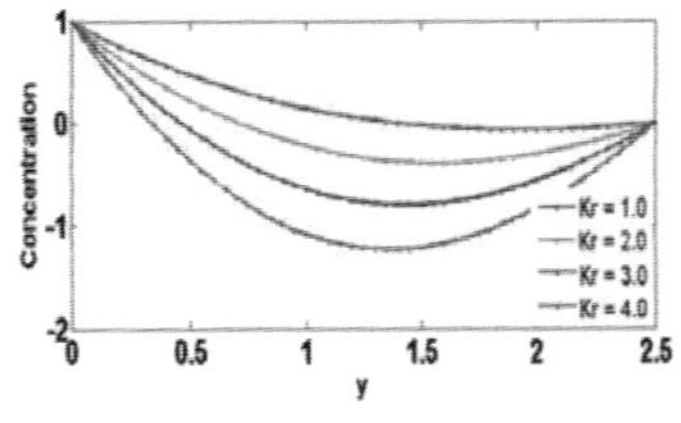

Figure 8. Variation of concentration with (K_r).

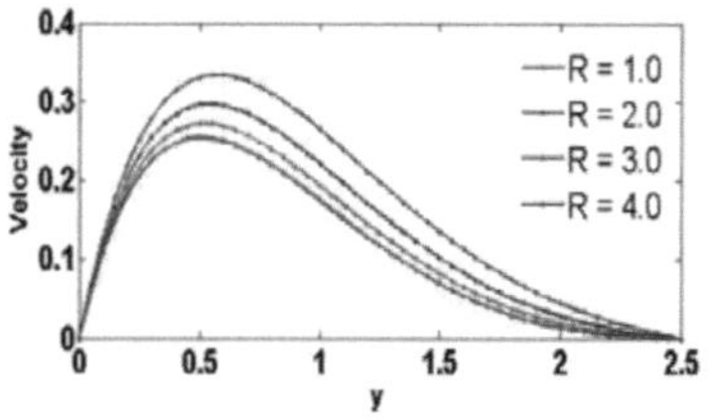

Figure 9. Variation of velocity with (R).

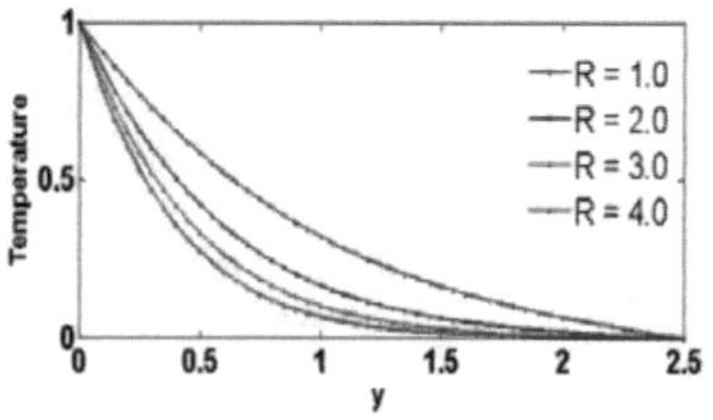

Figure 10. Variation of temperature with (R).

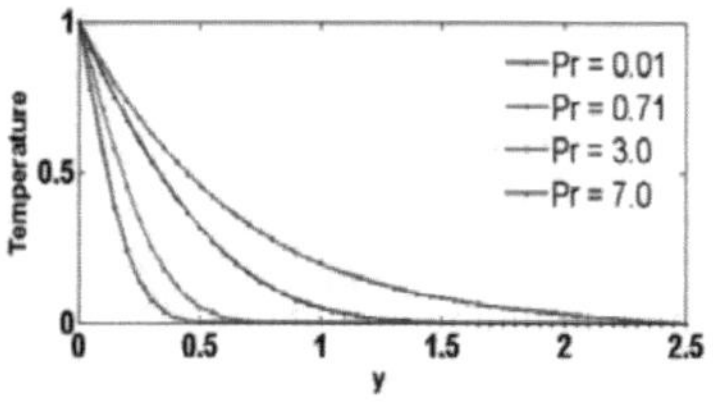

Figure 11. Variation of temperature with $(\Pr)$.

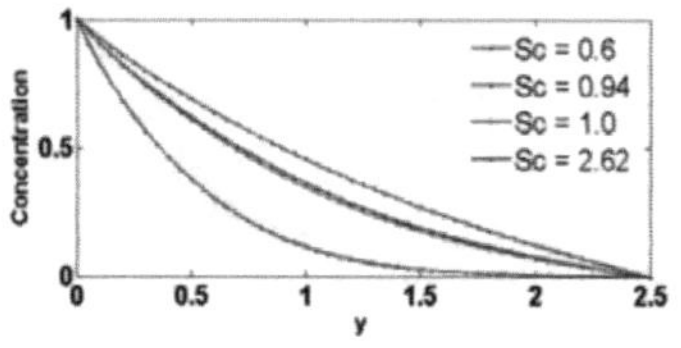

Figure 12. Variation of concentration with (Sc).

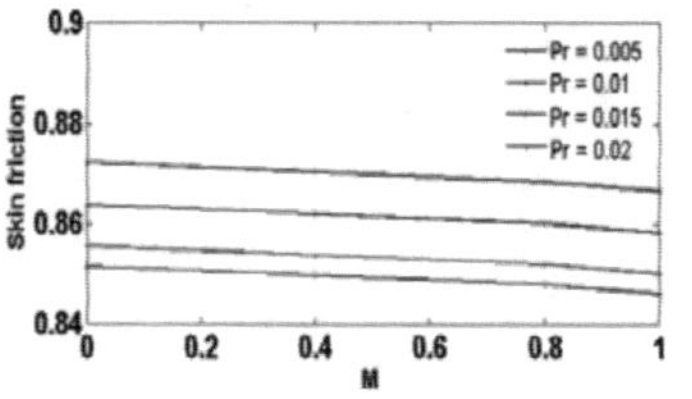

Figure 13. Variation of skin friction with (M & Pr).

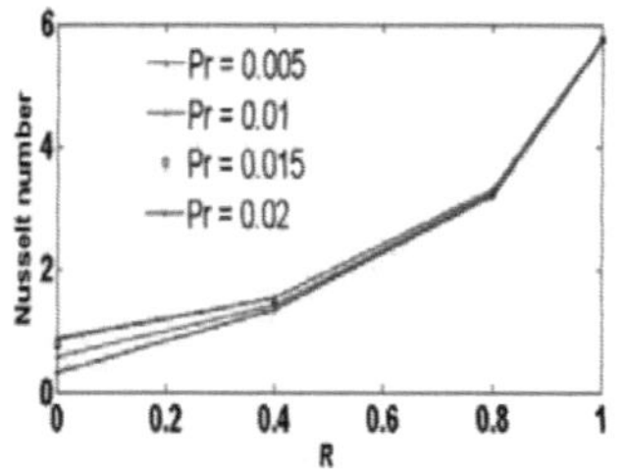

Figure 14. Variation of Nusselt number with (R & Pr).

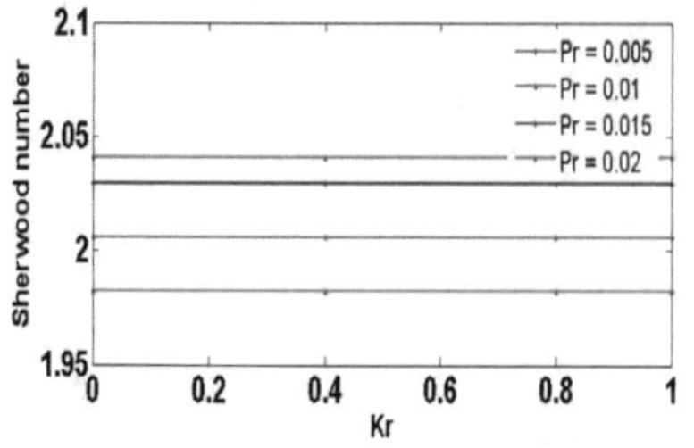

Figure 15. Variation of Sherwood number with (K_r & Pr).

5 Conclusions

This paper investigates the effects of radiation and chemical reaction on unsteady MHD free convection flow past an infinite vertical plate with variable suction. The governing partial differential equations are solved by an implicit finite difference method of Crank Nicolson type. The conclusions from the present study are as below: Velocity profile increases with increasing γ. It is also interesting to note that increase in M, R, K_r and γ decelerates the flow velocity. An increasing $\Pr, R, K_r$ and γ decrease the temperature profile of the fluid flow field at all points. Concentration of the fluid decreases with the increasing values of γ, Sc and K_r .Skin friction and Nusselt number tend to decrease rapidly as the Pr is increased, while the Sherwood number increases with increasing Pr.

This type of flow can be investigated in different geometries, like cylindrical coordinates, Couette flow etc. and effects of Navier slip on the flow system.

C – Concentration
C_p – Specific heat at constant pressure
D - Mass diffusivity
g – Acceleration due to gravity
Gr – Grashof number
Gc – Solutal Grashof number
k – Porous parameter
Nu - Nusselt number
Pr – Prandtl number
Sc – Schmidt number
R - Radiation parameter
K_r - Chemical reaction parameter
n - Order of chemical reaction
T –Temperature
C_f -Skin friction
Sh - Sherwood number
u, v – velocities in the x and y-direction respectively
x, y – Cartesian coordinates along the plate and normal to it respectively
B_0 - Magnetic field of constant strength
M – Magnetic field parameter

Greek Letters

$\beta^{'}$ - Coefficient of expansion with concentration
β - Coefficient of thermal expansion
$\sigma_0^{'}$ - Stefan-Boltzmann constant
σ - Scalar electrical conductivity
ρ- Density of fluid
θ- Dimensionless temperature
ν - Kinematic viscosity
γ - Suction parameter
λ - Variable thermal conductivity

Subscript
∞ - condition at infinity

References

1. M. Acharya, G. C. Dash, L. P. Singh. Magnetic field effects on the free convection and mass transfer flow through porous medium with constant suction and constant heat flux, Int. J. of Pur. & Appl. Maths., **31**, 1 1-18 (2000)
2. A. J. Chamkha. Unsteady MHD convective heat and mass transfer past a semi-infinite vertical permeable moving plate with heat absorption, Int. J. of Eng. Sci., **42,** 217-230 (2004)
3. P. Chandrakala, S. R. Antony. Radiation effects on MHD flow past an impulsively started infinite vertical plate with variable temperature, Int. J. of Maths, **48**, 2 167-175 (2006)
4. H. M. Duwairi, R. A. Damesh. MHD natural convection heat transfer from radiative vertical porous surfaces, Heat Mass Trans., **31**, 243-248 (1996)
5. K. A. Helmy. MHD unsteady free convection flow past a vertical porous plate, ZAMM, **78**, 4, 225-270 (1998)
6. M. A. Hossain, H. S. Takhar. Radiation effect on mixed convection along a vertical plate with uniform surface temperature, Heat Mass Trans., **31**, 243-248 (1996)
7. R. Kandasamy, K. Perisamy, P. K. Sivagnana. Effects of chemical reaction, heat and mass transfer along a wedge with heat source and concentration in the presence of suction or injection, Int. J. of Heat and Mass Trans., **48**, 1388-1394 (2005)
8. A. M. Kawala, S. N. Odda. Numerical investigation of unsteady free convection on a vertical cylinder with variable heat and mass flux in the presence of chemical reactive species, Adv. Pur. Maths, **3**, 1A 183-189 (2013)
9. P. M. Kishore, N. V. Prasada, S. V. Varma, S. Vantarakamana. The effects of radiation and chemical reaction on unsteady MHD free convection flow of viscous fluid past an exponentially accelerated vertical plate, Int. J. Phys. and Math. Sci. **4**, 1 300-317 (2013)
10. A. Mahdy. Effect of chemical reaction and heat generation/absorption on double-diffusive convection from a vertical truncated cone in a porous media with variable viscosity, Int. Comm. Heat and Mass Trans., **37**, 5 548-554 (2010)
11. F. Mohammad, O. Masatiro, S. Abdus, A. Mohamud. Similarity solution for MHD flow through vertical porous plate with suction, J. Comp. and Appl. Mech. **6**, 1 15-25 (2005)
12. P. M. Patil, A. J. Chamkha. Heat and mass transfer from mixed convection flow of polar fluid along a plate in porous media with chemical reaction, Int. J. Num. Meth. Heat Fluid Flow, 23, 5, 899-926 (2013)
13. R. Rajeshwari, B. Jothiram, V. K. Nelson. Chemical reaction, heat and mass transfer on non linear MHD boundary layer flow through a vertical porous plate in the presence of suction, Appl. Math. Sci. **3**, 50 2469-2480 (2009)
14. U. S. Rajput, S. Kumar. Radiation effects on MHD flow past an impulsively started vertical plate with variable heat and mass transfer, J. Appl. Mathe. Mech. **8**, 66-85 (2012)
15. A. Raptis, C. Perdikis. Radiation and free convection flow past a moving plate, Appl. Mech. Eng. **44**, 817-821 (1999)
16. M. A. Saleh, A. A. Mohamed, S. E. Mahmoud. Heat and mass transfer in MHD visco-elastic fluid flow through porous medium over a stretching sheet with chemical reaction, J. Appl. Maths. **1**, 446-455 (2010)
17. Sarada, Shankar. Chemical reaction effect on an unsteady MHD convective flow past an infinite vertical moving plate embedded in a porous medium with heat source, J. Glo. Res. Math. Arch. **1**, 5 63-81 (2013)
18. M. A. Sattar. Free and forced convection boundary layer flow through a porous medium with large suction, Int. J. En. Res. **17**, 1-7 (1993)
19. V. M. Soundalgekar, H. S. Takher. Radioactive convective flow past a semi-infinite vertical plate, Mod. Meas. Cont. **51,** 31-40 (1992)
20. M. M. Rahman. Combined effects of internal heat generation and higher order chemical reaction on the non-Darcian forced convective flow of a viscous incompressible fluid with variable viscosity and thermal conductivity over a stretching surface embedded in a porous medium, Can. J. Chem. Eng. **90**, 1631-1644 (2012)

Heavy metal concentrations of selected public parks of Istanbul City

Goksel Demir[1], H. Kurtulus Ozcan[2], Huseyin Ozdemir[3], Ali Osman Pektas[3], Ilker Oruc[4] *and* Muhammet Buyukyildiz[5]

[1]*Kirklareli University, Architecture Faculty, Department of Urban and Regional Planning, Kirklareli- Turkey*
[2]*Istanbul University, Engineering Faculty, Department of Environ. Engineering, 34320 Avcilar, Istanbul-Turkey*
[3]*Bahcesehir University, Faculty of Engineering, Department of Environ. Engineering, 34349 Besiktas, Istanbul-Turkey*
[4]*Kirklareli University, Vocational College of Technical Sciences, Kirklareli-Turkey*
[5]*Istanbul Grater Municipality Quality Control Laboratory, Istanbul-Turkey*

Abstract. Many cities, especially larger metropolises, parks are very important recreational areas where people usually have closer contact with flora. Therefore, the pollution level in the parks can have a greater effect on human health. Heavy metals are ubiquitous with the environment, as a result of both natural and anthropogenic activities, and humans are exposed to them through various pathways. Essentially, these areas are assumed to be less exposed to routine contaminants, but especially in metropolises, this assumption could prove false considering these areas are stuck within the confines of a city full of pollutant activity such as intense traffic. In this study; the relationships between heavy metal pollution levels (Cd, Cr, Cu, Ni, Pb, Zn) and the pH and electrical conductivity (EC) of soil samples were investigated from the parks on the Asian side of Istanbul. For this purpose, the most frequently visited 16 parks were selected as sampling sites. In the second part of the study, linear correlation is used for the data analysis.

1 Introduction

In addition to air and water, living organisms also need soil to survive. Land is an important environmental component for the future of living organisms. Unfortunately, the land is contaminated by different sources gradually. One of the most significant types of pollution on land is heavy metal pollution, thus this subject has become a focus point of environmental science. Among all of the researched heavy metal pollutants, Cu, Cd, Ni, Zn, Co and Pb become the most frequently examined, as they have a wide range of applications [1, 2, 3, 4]. Because of crop production over contaminated soils and contaminated pastures, heavy metals are carried up the food chain, and this process is therefore significantly affecting all living organisms in the ecosystem – with human beings as no exception – in a dangerous manner [5].

The earth on which we live and produce our food is polluted for many reasons, such as irresponsible agricultural methods, mining operations, residential operations and industrial applications of methods in agriculture. When the soil is polluted by the wastewater, this situation can lead dangerous consequences via groundwater contamination. The most important and known effect of soil pollution by the perspective of environmental health is the mobility of pollutants from soil to plants and then from plants to animals and people. Additionally, there are many health risks associated with physical exposure to contaminated soil, or digestion of contaminated dust – especially digestion of evaporating mercury and arsenic, in the soil [6]. Some important physical and chemical properties of soil, such as texture, cation exchange capacity, pH value and the amount of organic matter within the soil, are important parameters that affect the heavy metal accumulation rate of the soil in question [7]. Heavy textured soils are particularly able to absorb high ratios of heavy metal since they have a high exchange capacity of cation. Furthermore, soils that are rich in organic matter can much more quickly absorb heavy metals and cause the formation of hard-soluble compound.

Heavy metals are spread into the atmosphere from many different sources and different processing stages. In particular, heavy metals released from various sources, such as petrochemical, paper, energy, transportation and fertilizer industries, can cause damage to the ecological balance by following the path of dry and wet deposition to the surface and surface waters, and then mixed up in groundwater [2-8]. Generally, soil is able to neutralize the high level elements in it and so prevent aquatic organisms from the toxic effects of these elements. But, there could be heavy metal contamination close to the domestic and industrial waste places, heavy traffic roads and the unconsciously fertilized agricultural areas [9]. Common activities that increase the heavy metal accumulation of soil include the use of heavy metal-containing fertilizers and pesticides in agricultural activities and landscape studies, using high content heavy metal surface waters for irrigation and the use of heavy metal contaminated sewage sludge and wastes in order to increase the soil productivity. Another increasing factor of soil heavy metal accumulation is rainwater that has been polluted by flue and exhaust gas emissions [10].

In literature, the heavy metal studies that focused on recreational areas and parks are rare. Kabala et al. [11] examined total concentrations of Cu, Pb and Zn from 180 small family gardens of Poland and found that up to 35% of soils are excessively contaminated and require reclamation. Rajan et al. [12] investigated the heavy metal chemical changes before and after a monsoon at the recreational park tributaries in Pahang, Malaysia. They took water samples from 5 tributaries 6 times (3 times before, and 3 after the monsoon) and monitored the temperature, pH, oxygen derivation and the quantities of 8 heavy metals (Ag, Cd, Cu, Cr, Fe, Ni, Pb, and Zn). They too, also concluded that the water supply in the investigated National Parks have acceptable quality for recreational purposes. On the other hand, the usage of multivariate statistical analysis in heavy metal studies is very common. Arslan et al. [13] used cluster analysis and principal component analysis to comprehend the relationships between salinity, sodicity, and some soil properties in irrigated areas of the Bafra Plain, Turkey. They took 86 samples from different depths of 0-30 cm and 30-60 cm and used cluster analysis to group the sampling sites. Similarly, Liu et al. [14] used principal component analysis and cluster analysis to evaluate the heavy metal concentrations in soils around two gauge hills from Zhuxiangzhuang coal mine, northern Anhui province, China. Shahbaz et al. [15] analyzed the egg contents, prey and soil samples of little egret and cattle egret from two headworks to assess the importance of prey and habitat contamination as an exposure source for heavy metals. They used Correlation Analysis and Hierarchical Agglomerative Cluster Analysis (HACA) and identified relatively similar associations of metals and their source identification. Zheng et al. [16] used multivariate statistical analysis (correlation analysis, principal-components analysis, and clustering analysis) to investigate the heavy metal content of particulate matter in the city of Guangzhou in southern China by using samples of urban foliage near 36 pedestrian bridges.

The purpose of this study is investigating the relationship between heavy metal pollution levels (Cd, Cr, Cu, Ni, Pb, Zn) and the pH and electrical conductivity (EC) of soil samples from the parks on the Asian side of Istanbul. Linear correlation analysis is used for this purpose.

2 Material and method

2.1 Sampling procedure

This study was conducted in the city of Istanbul, the most important city in Turkey in terms of its industry and economy. Beside the intensive industrial facilities and economic mobility of the city, the high population intensifies the problems of environmental pollution. Istanbul is an intercontinental city, which is between Europe and Asia. The European part of the city is called the European side and the Asian part is called the Asian Side. According to a report which is prepared by the Turkish Statistical Institute (TSI), the city has a total population of 13.6 million people and 64.7% of this population is living in the European side while the other 35.3% living in the Anatolian side [17]. Since most of the parks are isolated places when compared with general urban areas, it was intended to select the most probably contaminated parks in order to show the relationship of heavy metal concentrations. For this purpose, the most frequently visited 16 parks were selected as sampling sites. The selected sampling sites (parks) are in the vicinity of the intense urban activity. As a result of this condition, selected parks are also near intense traffic roads and industrial activities. The texture of soils which were measured in this study is generally sandy, clayey, loamy, slight alkali, saltless, and contains low amounts of lime and medium amounts of organical material [18]. In the site selection phase of the study, this situation is taken into consideration and an assumption is made that these areas should be much more exposed to heavy metal contaminations, especially sourcing from the air and flue gas exhaust emissions. For sampling areas, another default pollutant factor is annual maintenance works such as fertilization and soil regulation processing works. After determining the sample sites, the soil samples were taken from two levels of depths (0-20 cm and 20-40 cm) by considering soil color, the slope, plant diversity and the difference of activities on the ground. For each determined sampling point, a sufficient number of soil samples were taken to that rep-resents corresponding parcel. Afterwards, these soil samples were mixed to obtain homogenous composite, and at least one kilogram of the sample was taken from each homogenous composition. The sampling points and sampling numbers are given in Table 1.

Table 1. Sampling point notation and the number of sampling.

Sampling points	Notation	Number of soil samples	
		0-20 cm	20-40 cm
Kadıköy Moda Parkı	P1	23	23
Göztepe Parkı	P2	11	11
Osman Gazi Korusu	P3	23	23
İdealtepe Parkı	P4	14	14
Gözdağı Parkı	P5	17	17
Pendik Üst Kaynarca	P6	15	15
Tuzla Botaş Parkı	P7	18	18
Büyükçamlıca Parkı	P8	16	16
Küçükçamlıca Parkı	P9	20	20
Millet Parkı	P10	6	6
Beykoz Korusu	P11	15	15
Beykoz Çayırı	P12	10	10
Hidiv Kasrı	P13	9	9
Fethipaşa Korusu	P14	11	11
Doğancılar Parkı	P15	4	4
Üsküdar Sahil Parkı	P16	7	7
Total		**219**	**219**

2.2 Sampling analysis

When Heavy metal analysis based on the ISO 11464 [19] standards, soil samples were subjected to a drying process in an air circulated oven in which the loss of moisture or 24-hour did not exceed 5% by weight (on average 35 - 40 °C). To ensure homogeneity, soil samples have been prepared for analysis by grinding the samples with wood and porcelain mortar and hammers until the particle diameter is convenient to pass under a 12:18 mm sieve. Wood and porcelain mortars and hammers were used to prevent from metal contamination. Then, 0.2-0.3 g weighing analysis samples were taken over teflon vessels from this homogenized soil samples, and a 10 ml acidic solution (1:3 HNO_3: HCl) was added on this analysis sample. After a one hour digestion process under a Microwave power (Berghof Speed), the resulting extract was then diluted with distilled water at the desired rate and has been made ready for measurements. The heavy metal (Cd, Cr, Cu, Ni, Pb, and Zn) concentrations were measured by using the ICP-OES instrument (Perkin Elmer, Optima 2100 [20].

2.3 pH and Electrical Conductivity analysis

To measure the pH values of soil samples first, 5ml soil analysis samplings have taken and 25 ml of pure water was added on to obtain a suspension. Subsequently, this suspension was stirred vigorously and has been waited from 2 hour to 18 hour to access stabilization period of suspension. After stabilization, the pH was measured by using an appropriate instrument (WTW Multi 340i) on the filtrated suspension. The electrical conductivity of the soil was also measured with the same instrument. Before measurement 20 g of analysis samples were taken from a homogenized soil mixture. Pure water is used to suspensionized these samples, then suspensions are shaken for 30 minutes at 180 rpm agitation rate value. After agitation, the suspension was filtrated and the Electrical Conductivity (EC) values were measured with the instrument WTW Multi 340i.

3 Result and discussion

3.1 Heavy metals of soil samples

The mean values of heavy metal concentrations for each sampling point (parks) are presented in Table 2 and 3 related to the soil sample depths as 0-20 cm and 20-40 cm, respectively. In nature; cadmium and zinc are associated with one another. Cadmium is obtained as a by-product during the refination of zinc.

Table 2. The heavy metal concentrations of samples in the depth range of 0-20 cm

	pH	*EC*	*Cd*	*Cr*	*Cu*	*Ni*	*Pb*	*Zn*
		µmho s/cm	*mg/ kg*	*mg/ kg*	*mg/ kg*	*mg/ kg*	*mg/k g*	*mg/k g*
P1	7.8	121.1	0.3	45.	27.	26.	14.1	63.4
	30	30	10	810	780	810	50	00
P2	7.7	118.3	0.3	54.	19.	35.	12.3	82.3
	90	60	90	530	760	560	40	20
P3	7.2	123.3	1.1	47.	38.	30.	35.5	68.1
	90	20	40	810	260	820	20	60
P4	7.6	230.9	0.3	52.	38.	33.	15.6	69.6
	70	30	20	770	600	820	10	60
P5	7.6	142.0	2.6	44.	37.	29.	154.	189.
	40	00	70	890	180	500	420	690
P6	8.0	159.2	0.8	48.	37.	37.	28.3	115.
	40	00	90	630	110	210	90	480
P7	7.7	188.4	0.3	49.	21.	38.	19.7	53.5
	30	40	20	240	430	910	20	10
P8	6.6	146.9	0.4	71.	31.	31.	26.8	63.9
	50	40	20	960	490	740	80	30
P9	7.4	149.4	0.2	53.	27.	30.	24.5	55.7
	20	00	70	050	880	250	60	70
P1	7.9	266.6	0.1	42.	35.	34.	20.7	61.9
0	40	70	40	620	610	570	10	30
P1	7.1	127.3	2.5	66.	50.	64.	33.0	118.
1	10	30	40	270	830	310	20	690
P1	7.5	228.2	1.5	50.	50.	40.	28.1	101.
2	30	00	60	260	700	980	10	180
P1	6.5	148.3	1.7	68.	45.	54.	36.9	107.
3	50	30	60	040	640	130	50	110
P1	7.2	160.1	1.4	83.	53.	59.	36.3	111.
4	40	80	80	540	450	090	60	290
P1	7.6	204.2	1.5	51.	54.	44.	112.	129.
5	40	50	40	040	400	510	450	100
P1	7.6	257.7	1.1	44.	30.	35.	24.4	79.4
6	30	10	10	160	550	400	60	30
me	*7.4*	*173.2*	*1.0*	*54.*	*37.*	*39.*	*38.9*	*91.9*
an	*81*	*74*	*53*	*664*	*542*	*226*	*78*	*16*
mi	*6.5*	*118.3*	*0.1*	*42.*	*19.*	*26.*	*12.3*	*53.5*
n.	*50*	*60*	*40*	*620*	*760*	*810*	*40*	*10*
ma	*8.0*	*266.6*	*2.6*	*83.*	*54.*	*64.*	*154.*	*189.*
x.	*40*	*70*	*70*	*540*	*640*	*310*	*420*	*690*
SD	*0.4*	*49.63*	*0.8*	*11.*	*10.*	*11.*	*38.4*	*36.0*
	24	*6*	*14*	*625*	*979*	*003*	*32*	*47*

Cadmium compounds are used as stabilizers or pigments in; the coating process of metals, alkali coils, several metal alloys such as copper and in plastics. Because of this, the cadmium concentration is at high levels where the places where industrial activities are intense

Table 3. The heavy metal concentrations of samples in the depth range of 20-40 cm

	pH	*EC*	*Cd*	*Cr*	*Cu*	*Ni*	*Pb*	*Zn*
		µmho s/cm	*mg/ kg*	*mg/ kg*	*mg/ kg*	*mg/ kg*	*mg/k g*	*mg/k g*
P1	8.1	161.3	0.3	52.	33.	32.	18.0	68.7
	90	00	70	510	920	860	60	20
P2	8.0	122.2	0.3	58.	19.	37.	14.7	71.5
	00	70	50	190	170	130	60	00
P3	7.2	126.4	1.0	40.	38.	28.	25.7	62.0
	40	20	40	500	570	930	40	10
P4	7.6	203.3	0.2	46.	33.	30.	11.5	61.2
	90	60	50	560	610	500	70	20
P5	7.6	115.8	2.7	43.	31.	29.	142.	173.
	60	00	80	640	580	470	220	860
P6	8.1	161.2	0.8	46.	37.	35.	24.7	122.
	30	70	90	340	090	030	10	360
P7	8.0	161.6	0.3	48.	23.	35.	28.2	58.9
	40	70	90	280	390	890	30	90
P8	6.7	153.7	0.3	68.	31.	43.	23.5	57.9
	00	50	50	860	490	080	50	20
P9	7.3	135.5	0.2	45.	28.	25.	25.4	51.1
	60	50	60	610	730	280	40	10

P1	7.9	250.5	0.1	43.	31.	29.	25.2	54.2
0	90	00	20	020	850	700	60	80
P1	7.1	129.1	2.4	59.	48.	58.	30.2	115.
1	20	30	90	380	670	010	10	540
P1	7.8	213.7	1.6	44.	45.	38.	23.4	93.7
2	90	00	30	930	790	500	00	80
P1	7.0	58.78	2.2	68.	48.	63.	26.9	94.2
3	10	0	10	830	560	680	20	50
P1	7.2	166.1	1.5	76.	53.	61.	33.8	110.
4	70	80	10	430	360	250	00	070
P1	7.9	183.0	1.6	55.	60.	45.	120.	131.
5	10	00	60	710	680	770	690	880
P1	7.9	195.7	1.1	58.	30.	37.	30.0	78.0
6	90	10	80	900	400	830	90	30
me	*7.6*	*158.6*	*1.0*	*53.*	*37.*	*39.*	*37.7*	*87.8*
an	*37*	*49*	*92*	*606*	*304*	*557*	*91*	*45*
mi	*6.7*	*58.78*	*0.1*	*40.*	*19.*	*25.*	*11.5*	*51.1*
n.	*00*	*0*	*20*	*500*	*170*	*280*	*70*	*10*
ma	*8.1*	*250.5*	*2.7*	*76.*	*60.*	*63.*	*142.*	*173.*
x.	*90*	*00*	*80*	*430*	*680*	*680*	*220*	*860*
SD	*0.4*	*45.54*	*0.8*	*10.*	*11.*	*11.*	*37.2*	*34.8*
	57	*3*	*69*	*728*	*251*	*917*	*04*	*09*

As shown in Table 2, for all parks and at the depth between 0-20 cm, the overall mean values of cadmium concentration were 1.050 mg/kg, whereas, the minimum and the maximum values appeared to be 0.140 mg/kg and 2.670 mg/kg at the parks P10 and P5, respectively. Similarly, the mean cadmium concentration is 1.09 mg/kg at the depth level 20-40 cm (Table 3) and the min-max Cd concentration are observed in P10 and P5 parks, respectively. The mean Zn concentration of 0-20 cm depth is 91.920 mg/kg where the maximum Zn concentrations were measured from the samples of P5 for both depth ranges. For 0-20 cm depths the minimum Zn concentration is observed at P7 and in the depth range of 20-40 cm the minimum concentration of Zn may be observed in P9 (Table 2-3).

Mostly, the chromium exists in the form of insoluble chromic acid $Cr_2O_3H_20$ in soil and this form is not mobile in soil structure. The washing process of soil via irrigation and precipitation increases the mobility of chromium depending on whether the chromium has formed soluble compounds or not. As seen in Table 2, the overall mean value of Cr is 54.664 mg/kg. The minimum and maximum concentrations of Cr are observed in P10 and P14, respectively. For the depth level 20-40 cm, the overall mean value is calculated as 53.606 mg/kg, whilst the minimum and maximum concentrations are found in P3 and P14, respectively. The overall mean concentrations of Cu are 37.542 mg/kg and 37.304 mg/kg for the level depths of 0-20 cm and 20-40 cm, respectively. For both levels, the minimum and the maximum concentrations are measured in the parks P2 and P15. The Ni concentrations mean values vary from 39.226 mg/kg to 39.557 mg/kg with changing depth levels from 0-20 cm to 20-40 cm. In near surface levels, the minimum and maximum Ni concentrations are observed in P1 and P11, and for deep level samples minimum and maximum levels are measured in the samples of the P9 and P13, respectively.

In this study; another investigated heavy metal compound was Pb. The majority of the atmospheric Pb is revealed as a result of the combustion of alkyl-lead that has been used as an additive in fuels. Pb emissions released into the atmosphere is condensed in the soil thanks to precipitation, and then subsequently becomes party to the entire food chain via plants and animals. The Pb mean concentrations of the parks are 38.978 mg/kg and 37.791 mg/kg for shallow and deep levels, respectively. The minimum and maximum Pb concentrations are observed in P2 and P5 for 0-20 cm depths, whereas these values are observed in the parks P4 and P5 for the depth level of 20-40 cm (Table 2-3). Limit values for related heavy metals in national soil pollution control regulation were given in Table 4.

Table 4. Sampling point notation and the number of sampling.

Pollutant	Limit value for skin exposure (mg/kg dry soil)	Limit value for pollutant transportation to ground water (mg/kg dry soil)
Pb	400	135
Cd	70	27
Cr	235	10
Cu	3129	514
Ni	1564	13
Zn	23464	6811

As seen in Table 4, Pb, Cd, Cr, Cu, Zn concentrations of measured samples are lower than national soil pollution regulations. But Ni concentrations of samples are higher than limit value for pollutant transportation to groundwater contamination. Heavy metal concentrations some different regions were adopted from Elmaslar et.al, [21] and given in Table 5. When we compare the average heavy metal values determined in soil samples of this study with the obtained results in the literature, it can be noted that heavy metal concentrations are similar with results obtained from the other studies.

Table 5. Heavy metal concentrations of soil samples in different studies

Location	**Cu** *(mg/kg)*	**Ni** *(mg/kg)*	**Zn** *(mg/kg)*	**Pb** *(mg/kg)*	**Cd** *(mg/kg)*
Industrial area [22]	21.520	-	42.710	0.380	0.121
Agricultiral soil (0-30 cm), polluted area, Bulgaria [23]	58.300	-	338.000	217.00	3.590
Industrial area, Hong Kong [24]	26.100	-	62.80	87.700	1.310
Agricultiral soil, Manisa, Turkey [25]	14.300	22.300	6.750	27.300	2.800
Agricultiral soil, Menemen,	14.400	42.700	5.620	41.000	2.700

Turkey [25]					
Top soil, Istanbul City [21]	30.500	44.500	207.600	17.700	0.460

3.2 pH and electrical conductivity analyses

The results of pH concentrations and EC values are presented in Table 2 and 3 for the 16 parks on the Asian side of Istanbul, with regard to different sampling depths. For the activities of soil organisms and plant-availability of nutrients, the optimum pH level should be between 6-7 values. While plant growth is possible both below and above these pH values, there could be some disadvantages, therefore the soil nutrient deficiencies or potential toxicity could be predicted by observing the pH levels. As shown in Table 2 and Table 3, the pH values of soil samples vary within the scale of neutral (6.5-7.5) and slightly alkaline (7.5-8.5).

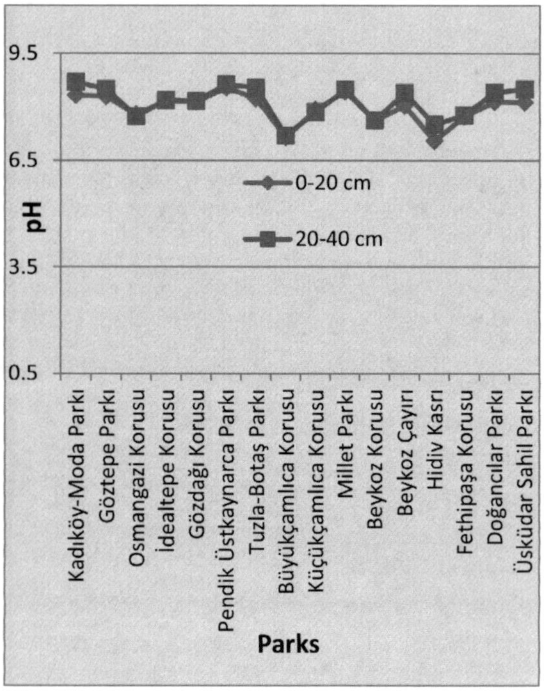

Figure 1. Variation of the pH values according to their locations and depths.

While the soil depths change between two levels, the pH levels appear to be stable regardless, i.e. there are no serious differences between pH values (Fig. 1). As is well known, organic and chemical fertilizers that are used as manure usually cause the increase of salt concentrations in soil, since they consist of dissolved materials. While the increase in concentration levels affects the osmotic pressure, this situation leads a dehydrational media where even if there is adequate water, the plants can not benefit. Additionally, a negative consequence of this situation is that the water within the plant can filtrate through the ground. As shown in Table 2 and 3 the investigated soil samples have a mean EC value of 173.274 μmhos/cm and 158. 649 μmhos/cm for shallow and deep data set, respectively. As shown in Fig. 2, EC values do not show significant differences except Hidiv Kasri. By these results it can be asserted that all soil samples belong to the salt-free soil class (0-4000 μmhos/cm) by the sense of the EC.

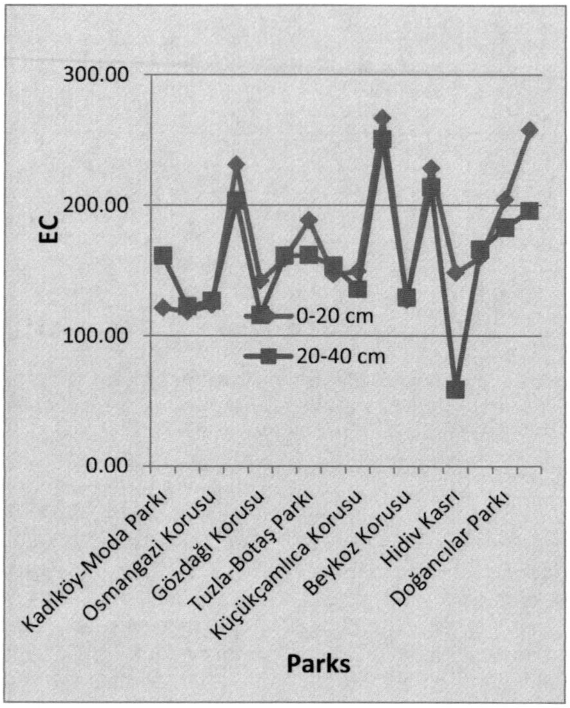

Figure 2. Variation of the EC results (μmhos/cm) according to their locations and depths.

3.3 Determination of correlation between heavy metals, pH and EC

Correlation analysis method has been applied to the data set that is obtained from the experimental results of soil samples. The existence of a linear relationship between heavy metal concentrations, pH and EC values, were investigated. At this stage of statistical analysis, it was assumed that there would not be a mutual interaction between variables and there existed only a linear relationship. For this purpose, the widely used parametrical Pearson Correlation Analysis was employed to investigate the way of interaction, whilst the determinant coefficients were used to examine the strength of the relationship.

The existence of a relationship between two variables, and the direction of the relationship, can be examined by correlation and determination analysis. In this analysis, it is not taken into consideration whether the variables are dependent or independent variables. In scientific literature, correlation coefficients are usually grouped into parametric and non-parametric correlation coefficients. The most well-known non parametric correlation coefficients are Spearmen and Kendall

DOI: 10.1051/matecconf/20166402001

coefficients. On the other hand, the most known and widely used parametric methods are the Pearson correlation coefficient [26]. The values of the Pearson correlation coefficient vary between -1 and +1, the coefficient is about the values of zero when there is no linear relationship, and the close values to absolute one indicate that there is a strength relationship between two variables. Pearson's correlation coefficient between two variables is defined as the covariance of the two variables divided by the product of their standard deviations [27]. It must be pointed out that the Pearson correlation coefficient does not determine causalities, nor ascertain the presence of nonlinear relations, however, only linear relationship [28]. The Pearson correlation coefficient matrix of the dataset and the significance values of the correlations are presented in Table 6. Different depth observations have been merged into a dataset since there is not a significant difference between the results of different depth values.

Table 6. Correlation coefficients of pH, EC and Heavy metal concentrations [29]

	pH	*EC*	*Cd*	*Cr*	*Cu*	*Ni*	*Pb*	*Zn*
pH	*1*	*0,377*	*-0,295*	*-0,611*	*-0,298*	*-0,451*	*0,037*	*-0,003*
Sig. (2-tailed)		*0,033*	*0,102*	*0*	*0,098*	*0,01*	*0,841*	*0,987*
EC	*0,377*	*1*	*-0,315*	*-0,319*	*0,06*	*-0,215*	*-0,098*	*-0,192*
Sig. (2-tailed)	*0,033*		*0,079*	*0,075*	*0,743*	*0,238*	*0,593*	*0,292*
Cd	*-0,295*	*-0,315*	*1*	*0,218*	*0,632*	*0,552*	*0,623*	*0,839*
Sig. (2-tailed)	*0,102*	*0,079*		*0,231*	*0*	*0,001*	*0*	*0*
Cr	*-0,611*	*-0,319*	*0,218*	*1*	*0,383*	*0,755*	*-0,12*	*0,087*
Sig. (2-tailed)	*0*	*0,075*	*0,231*		*0,03*	*0*	*0,513*	*0,638*
Cu	*-0,298*	*0,06*	*0,632*	*0,383*	*1*	*0,67*	*0,335*	*0,514*
Sig. (2-tailed)	*0,098*	*0,743*	*0*	*0,03*		*0*	*0,061*	*0,003*
Ni	*-0,451*	*-0,215*	*0,552*	*0,755*	*0,67*	*1*	*-0,011*	*0,317*
Sig. (2-tailed)	*0,01*	*0,238*	*0,001*	*0*	*0*		*0,953*	*0,077*
Pb	*0,037*	*-0,098*	*0,623*	*-0,12*	*0,335*	*-0,011*	*1*	*0,807*
Sig. (2-tailed)	*0,841*	*0,593*	*0*	*0,513*	*0,061*	*0,953*		*0*
Zn	*-0,003*	*-0,192*	*0,839*	*0,087*	*0,514*	*0,317*	*0,807*	*1*
Sig. (2-tailed)	*0,987*	*0,292*	*0*	*0,638*	*0,003*	*0,077*	*0*	

As seen in Table 6, the correlation coefficient between pH-Cd and pH-Cu is r= -0.295 and r=-0.298, respectively indicating that there is an inverse low correlation. Furthermore, there are inverse relationships between the pH and all investigated heavy metals with the exception of Pb. The lowest correlated parameters with pH are between Zn and Pb, with a correlation coefficient r=-0.003 and r=0.037, respectively. These values are close to zero and negligible, thus for this dataset it can be concluded that the pH level is not influenced by Zn and Pb accumulation in park soils. On the other hand, the highest correlated parameter is Cr. The correlation coefficient between pH and Cr is r= -0.611. This value is between the range of $0.5<|r|<0.9$ so it can be asserted that the power to influence the concentration of Cr by soil pH is at a moderate level. The correlation between pH and Ni is r=-0.451, which is in low correlation zone ($0.0<|r|<0.5$).

The correlation coefficients between EC and heavy metal concentrations vary between a very low range $0.06<|r|<0.319$, therefore it can be asserted that there is no significant relationships between EC values and heavy metal concentrations. From the perspective of heavy metal cross correlations Cd and Zn (r= 0.839), Cr and Ni (r= 0.755), Pb and Zn (r= 0.807) have significant correlations.

4 Conclusion

By the results of this study the pH values of 0-20 cm samples vary between to 6.5-8.1 ranges, indicating that the soils belong to neutral and slightly alkaline soil types. As is known, in plant cultivation studies the preferred soil pH values range from 6.5 - 7.5 since because of this range the plants could retrieve the needed nutrients easier. Since the pH values are over the edge value (>7.5) in most parks (P1, P2, P4, P5, P6, P7, P10, P12, P15 and P16) it would be useful to include acidic fertilizers or sulfur the applications into the seasonal maintenance programs by terms of plant nutrition.

By examining the relationships with using correlation analysis between the heavy metal levels and EC and pH values it can be concluded that there lies a relationship between pH, Cr and Ni levels. Similarly, there is a relationship between Cd, Pd and Zn levels. For this data set the EC is found unrelated to each of the examined element levels.

Having tested the various sites, heavy metal concentrations of the samples were found below the limit values that are set by the National Soil Pollution Control Regulation [30]. However, within such areas, heavy metals could accumulate in the soil surface or (and) could filtrate the soil over time and cause threatening levels of contamination. In order to prevent this situation, uncontrolled and excessive use of pesticides and chemical fertilizers, especially in seasonal maintenance programs, should be avoided.

Acknowledgments

This study, a part of master thesis entitled "The soil quality and heavy metal content analysis in some specified playing grounds of Istanbul Anatolian side" which conducted at Bahcesehir University Natural Science Institute.

References

1. G. Uyar, E. Avcil, M. Ören, F. Karaca, M.S. Öncel, Determination of Heavy Metal Pollution in Zonguldak (Turkey) by Moss Analysis (Hypnumcupressiforme) Environ Eng Sci **26**(1):183-194. doi: 10.1089/ees.2007.0253 (2009)
2. A. Sahmurova, H. Turkmenler, E.E. Özbas, Biosorption Kinetics and Isotherm Studies of Cd(II) by Dried Enteromorphacompressa Macroalgae Cells from Aqueous Solutions. Clean – Soil, Air, Water, 38(10):936–941. doi 10.1002/clen.201000108 (2010)
3. T.T.T. Duong, B.Y. Lee, Determining contamination level of heavy metals in road dust from busy traffic areas with different characteristics. J Environ Manage 92:554-562 (2011)
4. H. Zhu, X. Yuan, G. Zeng, M. Jiang, J. Liang, C. Zhang, J. Yin, H. Huang, Z. Liu, H. Jiang, Ecological risk assessment of heavy metals in sediments of Xia-wan Port based on modified potential ecological risk index. Trans Nonferrous Met Soc China 22:1470-1477. doi. 10.1016/S1003-6326(11)61343-5 (2012)
5. N. Çağlarırmak, Gıda güvenliğinin çevre kirliliği yönünden incelenmesi (in Turkish). 7th National Environmental Engineering Congress, 24-27 October 2007, İzmir, Turkey (2007)
6. S. Khan, Q. Cao, Y.M. Zheng, Y.Z. Huang, Y.G. Zhu, Health risks of heavy metals in contaminated soils and food crops irrigated with wastewater in Beijing, China. *Environ Pollut* 152:686-692 (2008)
7. F. Zeng, A. Shafaqat, H. Zhang, Y. Ouyang, B. Qiua, F. Wua, G. Zhang, The influence of pH and organic matter content in paddy soil on heavy metal availability and their uptake by rice plants. Environ Pollut **159**:84-91 (2011)
8. A.I. Amouei, Z. Yousefi, A.H. Mahvi, K. Naddafi, M. Tahmasbizadeh, Heavy metal concentrations in industrial, agricultural, and highway soils in northern Iran. Environmental Justice **5**(3):153-157. doi:10.1089/env.2011.0038 (2012)
9. O. Mikanova, Effects of heavy metals on some soil biological parameters, J Geochem Explor **88**:220-223 (2006)
10. O. Morton-Bermea, M.B. Alvarez, E.H. Gaso, N. Segovia, Heavy metal concentrations in surface soils from Mexico City. Bull Environ Contam Toxicol **68**(3): 383– 388. doi 10.1007/s001280265 (2002)
11. C. Kabala, C. Tadeusz, S. Leszek, Factors influencing the concentration of heavy metals in soils of allotment gardens in the city of Wroclaw, Poland. Fresenius Environ Bull **18**(7):1118-1124 (2009)
12. S. Rajan, N.N.M. Firdaus, M. Appukutty, K. Ramasamy, Effects of climate changes on dissolved heavy metal concentrations among recreational park tributaries in Pahang, Malaysia. *Biomed Res-India* 23(1):23-30 (2012)
13. H. Arslan, H. Gunal, M. Guler, Assessment the soil properties affecting salinity and sodicity of Bafra plain using multivariate statistical techniques. Carpat J Earth Env **8**(1):81-90 (2013)
14. X. Liu, L. Sun, S. Chen, Heavy metal pollution of soil around the gangue hill: a case study from Zhuxiangzhuang coal mine, northern Anhui Province, China. International Conference on Energy, Environment and Sustainable Development October 21-23, 2011, Shangai, China (2012)
15. M. Shahbaz, M.Z. Hashmi, R.N. Malik, A. Yasmin, Relationship between heavy metals concentrations in egret species, their environment and food chain differences from two Headworks of Pakistan. Chemosphere **93**:274–282 doi: 10.1016/j.chemosphere.2013.04.078 (2013)
16. Y. Zheng, Q. Gao, X. Wen, Multivariate statistical analysis of heavy metals in foliage dust near pedestrian bridges in Guangzhou, South China in 2009. Environ Earth Sci **70**(1):107-113. doi 10.1007/s12665-012-2107-z (2013)
17. TUIK, Turkish Statistical Institute, Official newsletter. No: 16. January (2012)
18. M. Buyukyildiz, The Soil Quality and Heavy Metal Content Analysis in Some Specified Playing Grounds of Istanbul Anatolian Side. Master Thesis, Bahcesehir University, Urban Systems and Transport Management Program (2013)
19. TS ISO 11464. Soil quality; pretreatment of samples for physicochemical analyses.
20. EPA, Test Methods for Evaluating Soil Waste. Microwave Assisted Acid Digestion of sediments, sludges, soils and oils, January.3051A-1 (1998)
21. E.E. Ozbas, H.K. Ozcan, N. Balkaya, G. Demir, C. Bayat, Heavy Metal Amounts in Soil and Sediments of Surface Water Sources in the Industrial Regions of Istanbul. J Residuals Sci Tech **4**(2):89-94 (2007)
22. B. Skrbic, and S. Cupic, Trace Metal Distribution in Surface Soils of Novi Sad and Bank Sediment of the Danube River. J. Environ. Sci. Health Part A. A39 (6)1547–1558 (2004)
23. P. Bojinova, B. Georgiev, I. Kabakchiev, V. Krasteva, L. Stanislavova, H. Tchuldjian, G. Welp, and G. Brummer, Harmonization of the Methods for the Investigation of Heavy Metal Pollution of Soils and the Standardization of the Assessment Criteria

DOI: 10.1051/matecconf/20166402001

for Soil Protection. Report No. UBA-FB 96-071. Terytze, K.N. ed., Poushkarov Institute of Soil Science and Agroecology, Sofia and Institut fur Bodenkunde: Bonn, (1996)

24. T.B. Chen, J.W.C. Wong, H.Y. Zhou, and M.H. Wong, Assessment of Trace Metal Distribution and Contamination in Surface Soils of Hong Kong. Environ. Pollut. **96**, 61–68, (1997)
25. O.L. Elmaci, S. Delibacak, M. Secer, and A. Bodur, Fertility Status, Trace Elements and Heavy Metal Pollution of Agricultural Land Irrigated from the Gediz River. , Int. J. Water. 2, Nos.2/3, (2002)
26. J.L. Myers, A.D. Well, Research Design and Statistical Analysis (2nd ed.), Lawrence Erlbaum, p. 508, ISBN 0-8058-4037-0 (2003)
27. C.L. Joseph, K.W. Kevin, L.Y. Wan, Solar radiation modeling using ANNs for different climates in China. Energy Convers Manage **49**:1080–1090. doi 0.1016/j.enconman.2007.09.021 (2008)
28. J. Duricic, T. Erdik, A.O. Pektaş, P.H.A.J.M. Van Gelder, Mean normalized force computation for different types of obstacles due to dam break using statistical techniques. Water **5**:560-577. doi:10.3390/w5020560 (2013)
29. SPSS Algorithm, IBM SPSS Statistics 20 Algorithms, Copyright IBM Corporation, 1989, USA (2011)
30. TKKY, National Regulations of soil pollution control Turkish Official Gazette, No 27605, Turkey (2010)

MATEC Web of Conferences **64**, 02002 (2016) DOI: 10.1051/matecconf/20166402002
ICCMP 2016

Performance of supercritical methanol in polyurethane degradation

Lu Liu, Lihua Tang, Zibin Zhu, Yanhui Ni and Yongqiang Wu

East China University of Science and Technology, Research Institute of Chemical Technology, Shanghai 200237

Abstract. Polyurethane is a group of block copolymer which is composed of diisocyanate, chain extender, and polyol, including polyurethane foam, polyurethane elastomer, waterborne polyurethane, etc. This research focused on thermoplastic polyurethane elastomer (TPU) which is formed with 4,4'-diphenylmethane diisocyanate(MDI), poly(1,4-butanediol-hexanedioic acid) diolpolyester(PBA) and extended with 1,4-butanediol(BDO). The degradation of TPU was carried out with the help of methanol as the supercritical solvent. The SEM of the reaction residues revealed the process of the depolymerisation. The products were measured by GC-MS and found out to be PBA, BDO and 4,4'-methylene diphenyl carbamate (MDC) which is the methylate of MDI. GC-FID, HPLC-UV and GPC were used to further analysis. The experimental results showed that supercritical methanol performed outstandingly in TPU recycling, it needed lower temperature and shorter time than regular methods. At 230℃/70min, over 90% raw materials of TPU could be recovered .

1 Introduction

Polyurethane (PU) is composed of diisocyanate, polyether polyol or polyester polyol, and chain extender [1]. The property and utilization of PU could be manipulated by varying the raw materials. Consequently, it is one of the most promising polymers. Its products include rubber, fibre, resin, tackiness agent, coating material and functional polymer.

Recent chemical recycling methods of PU contain pyrolysis, hydrolysis, glycolysis [2-4], aminolysis, and hydrogenation et al. Among the above methods, glycolysis has more advantages in industrial applications because of the broad adaptability and mild reaction conditions. However, the glycolysis only recovers the polyol from PU, and needs 2~5h to complete the degradation reaction. The products are complicated and difficult to separate.

Supercritical methanol(T_c=239 ℃, P_c=8.1MPa)has moderate critical condition, weak hydrogen-bond interaction, and high dielectric constant. It performs excellently in polymer dissolving and extracting [5, 6]. It's an important replacement of supercritical water in polymer recycling. Especially in the depolymerization of the condensation polymer (such as PET [7, 8], PC [9], PBT [10, 11]), supercritical methanol not only plays the role of the solvent, but also participates in the scission of the polymer chain as the reactor. Besides excellent transmission characteristic and dissolution characteristic, it also has splendid reaction characteristic.

In this work, thermolysis was compared with supercritical methanolysis, in order to investigate the feasibility and effectivity of the later method, study of TPU degradation behavior in sub-/supercritical methanol was carried out. By analysis and calculation, we found the maximum yield of MDC, PBA and BDO were 98%, 92% and 93%, respectively. Consequently, the utilization of supercritical methanol in TPU degradation and recycle was efficient and promising.

2 Experimental

2.1 Material and equipment

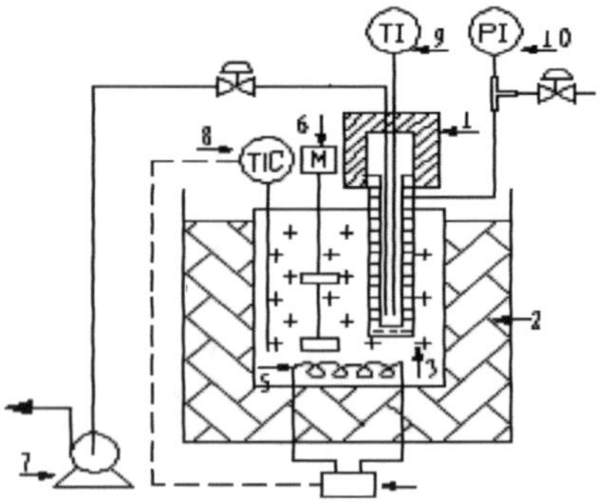

Figure 1. Schematic diagram of the experimental equipment
1.High pressure reactor 2.Insulating layer 3.Salt bath
4.Thermostat 5.Resistance wire
6.Stiring 7.Vacuum pump 8. Thermocouple
9.Temperature display instrument 10.Pressure transmitter

The polyurethane which was decomposed in this research was thermoplastic polyurethane elastomer (TPU): Bayer

DOI: 10.1051/matecconf/20166402002

Desmopan® CA205; The solvent was absolute methanol(AR,purity≥99.5%). The experimental apparatus is shown as Fig. 1, it was a high pressure batch reactor with a volume of 20ml made of 316L stainless steel. The reactor was heated by molten salts which contained 50wt.%KNO_3 and 50wt.%$NaNO_2$. The volume of the part exposed to the air is less than 0.1ml, so the affection of the unheated part to the whole system could be ignored. The temperature and pressure in the reactor were measured by pressure sensor and K-thermocouple, respectively.

2.2 Experiment and characterization

2.2.1 The feasibility of TPU degradation in supercritical methanol

1.0000g±0.0002g of raw TPU was weighed by electronic balance and fed into the above-mentioned high pressure batch reactor. The degradation condition was set as 250℃ /9.0MPa. The reactor was sealed up and vacuumized to -0.01Mpa by water ring vacuum pump. 12.5ml of methanol(the mass ratio of raw TPU and methanol is 1:10) was extracted by needle tubing and inhaled by the negative pressure. Then the reactor was heated by the melting salt. After 20min, the reaction was finished, the reactor was taken out of the molten salt and put into cold water immediately. When the temperature of the reactor reduced to ambient temperature, the degradation products were taken out and analyzed by GC-MS(GC: Agilent 6890, MS: Agilent 5973N).

In addition, A WRT-3P TG was used in testing 30mg of TPU in the atmosphere of N_2.

2.2.2 Degradation of TPU in sub-/supercritical methanol

The experimental procedure was the same as which was mentioned in 2.2.1. The degradation temperatures were in the range of 200~290℃, the corresponding pressures were 3.0~13.0MPa. The duration of the degradation was 5min～70min。

Analytical instruments : Haixin GC-930, FID; Waters600 HPLC, Waters486 UV(246nm) ; JEOL JSM-6360LV scanning electron microscopy(SEM).

2.2.3 Characterization of PBA in suitable degradation condition

At 230℃/7.2MPa, the degradation time was 5min～70min. After vacuum distillation, the products were dissolved in tetrahydrofuran(THF) and analyzed by GPC(PL-GPC50, mobile phase: THF) .

3 Results and discussion

3.1 The degradation results of supercritical methanol

The degradation products in supercritical methanol(250℃ /9.0MPa) were faint yellow liquid mixture. That indicated 20min was enough for supercritical methanol to degrade TPU completely. By contrast, TPU decomposition started at 300℃ and finished at 700℃ in thermolysis, as shown in Fig. 2. Therefore, supercritical methanol was superior to thermolysis on account of lower decomposition temperature and higher efficiency.

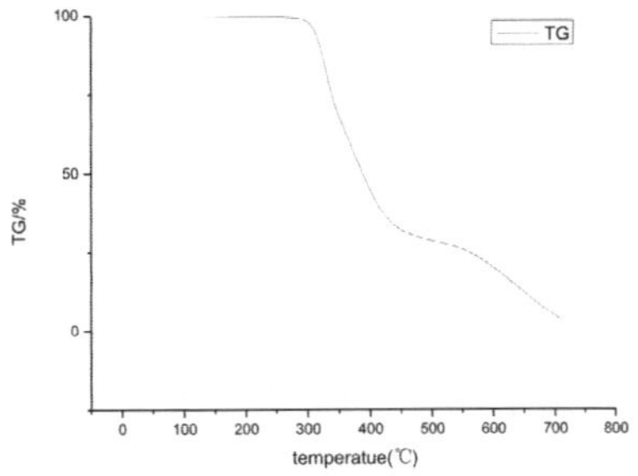

Figure 2. TG curve of TPU

Testing of the products mixture by GC/MS indicated that there were THF, BDO, dimethyl adipate (DMA), MDC, et al; Thereinto, THF was a byproduct from BDO cyclodehydration; A part of BDO was chain extender of TPU, the other was methanolysis product of PBA, as the same as BDO. GC was not proper to PBA which was a polymer itself. But from the methanolysis products of PBA, it could be inferred that the degradation products mixture also contained PBA.

3.2 Degradation of TPU in sub-/supercritical methanol

As the results shown above, the degradation products of TPU in supercritical methanol were MDC, BDO, PBA and its methanolysis products BDO, DMA. BDO and DMA were quantified by GC; MDC was quantified by HPLC [12]; PBA was measured by GPC. On the basis of these analyses, the influence of temperature and time on degradation rate and yields were obtained.

3.2.1 Analysis on degradation residue

The quality of the degradation residue was weighed and shown in Table 1. At 200℃/70min, the quality of the residue was 0.5076g. In the range of 200℃ to 240℃, the time of TPU decomposing completely was decreased significantly. When the temperature of the degradation was higher than 240 ℃ , the degradation reaction completed within 10min.

The SEM images of degradation residues in 200℃ to 250℃,10min were given as Fig. 3 to Fig. 9. There were plenty of micro areas on the surface of TPU, as shown in Fig. 3. Because TPU was a block copolymer consisting of hard segment and soft segment, hard segment was generated from MDI and chain extender, and could not be dissolved in soft segment which was composed of PBA. The alternative distribution of the two kinds of segments formed micro areas. This phenomenon was called microphase separation. In 200℃~210℃, the methanol molecule firstly infiltrated the amorphous soft segment. The polymer chains were decomposed to oligomers. As Fig. 4~Fig. 5 showed, the TPU fragments with microphase separation were observed. At 220 ℃, as shown in Fig. 6, the fragments continued to react with methanol , part of soft segment was dissolved in methanol, deprived from the polymer chains, the others twined around the hard segment. In Fig. 7~Fig. 8, the degradation temperature rising to 230℃~240℃, all the soft segments had removed from the TPU chains. Some of them attached to the hard segments which were spherical crystal arranged closely. At 250 ℃ /10min, supercritical methanol diffused into TPU, and the soft segments were rapidly ruptured from TPU . The internal structure of hard segments rearranged and congregated into flake-shaped fragment, as shown in Fig. 9.

Table 1. Quality of the degradation residue

Degradation	Quality of the degradation residue/g									
time/min	200℃	210℃	220℃	230℃	240℃	250℃	260℃	270℃	280℃	290℃
10	0.8084	0.7718	0.6043	0.3045	0.103	0.0173	0	0	0	0
20	0.7777	0.6812	0.4678	0.23	0	0	0	0	0	0
30	0.7323	0.5343	0.3291	0.1	0	0	0	0	0	0
40	0.6372	0.5025	0.1962	0.05	0	0	0	0	0	0
50	0.5769	0.3928	0.1433	0	0	0	0	0	0	0
60	0.5934	0.28	0.0827	0	0	0	0	0	0	0
70	0.5076	0.1514	0.0367	0	0	0	0	0	0	0

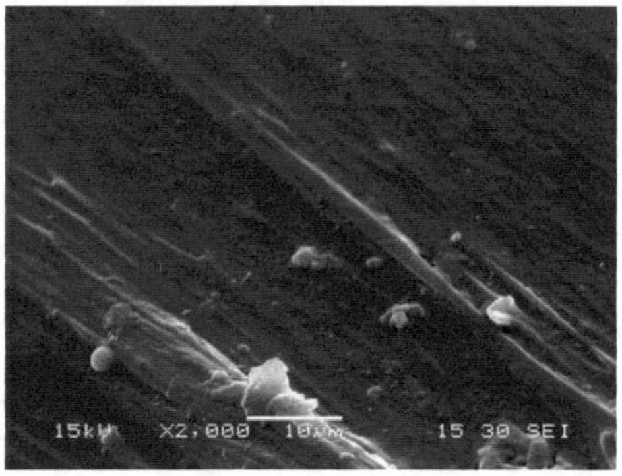

Figure 3. SEM image of TPU surface

Figure 5. SEM image of the degradation residue at 210℃ /10min

Figure 4. SEM image of the degradation residue at 200℃ /10min

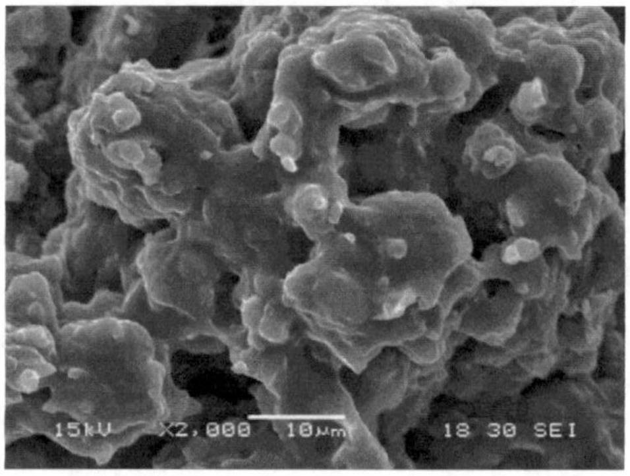

Figure 6. SEM image of the degradation residue at 220℃ /10min

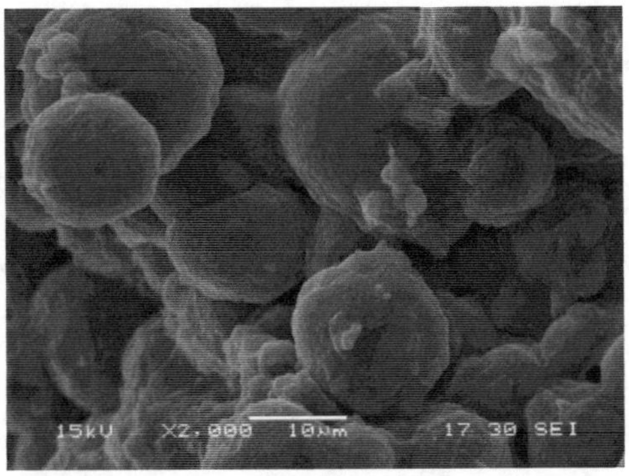

Figure 7. SEM image of the degradation residue at 230℃ /10min

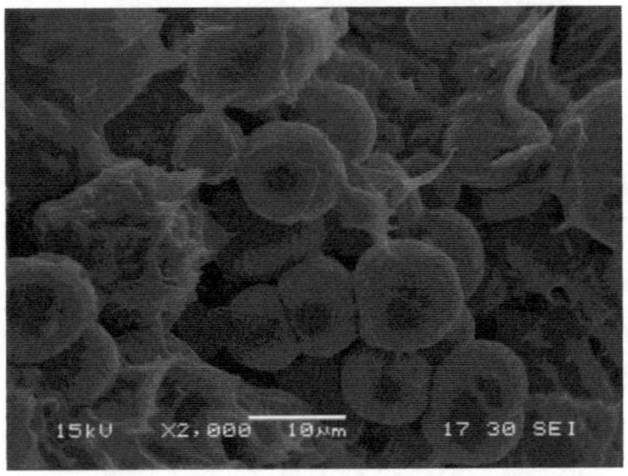

Figure 8. SEM image of the degradation residue at 240℃ /10min

Figure 9. SEM image of the degradation residue at 250℃ /10min

3.2.2 Analysis of the degradation products mixture

During the main products of the depolymerization, BDO, DMA were measured by GC internal standard method with naphthalene as the internal standard substance. While MDC had high boiling points, and would decompose above 200 ℃ . GC method was not appropriate.

To ensure the isocyanate(NCO⁻) reacting entirely, the products needed to be treated with excessive methanol before measuring by HPLC external standard method.

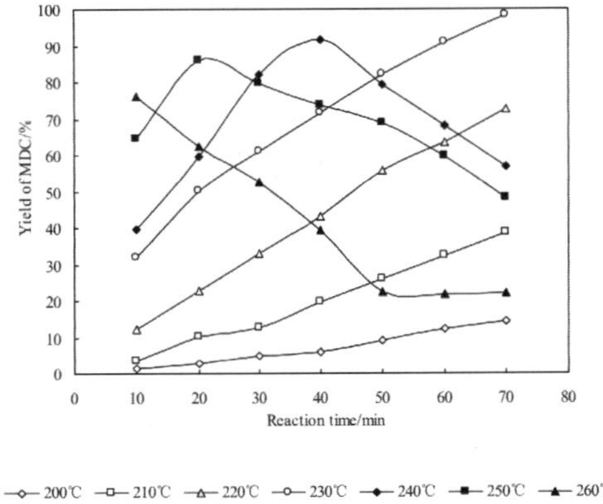

Figure 10. Relationship between yields of MDC and temperature

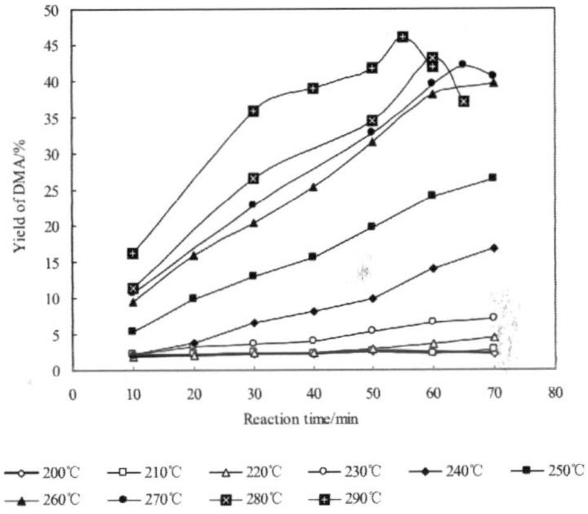

Figure 11. Relationship between yields of DMA and temperature

The influences of temperature and time on the main products MDC were represented in Fig. 10. In the temperature of 200~230℃, the yield of MDC increased with the temperature. In 240~260℃, the yield of MDC exhibited apparent peak values which showed earlier with increasing temperatures. On account of the secondary reaction above 250℃[12], the yield of MDC reduced sharply. The maximum yield of MDC reached 98% at 230℃ /70min. Higher temperature was not good for MDC recycling.

DMA and part of BDO were decomposed from PBA. The yield of DMA was low(approximate 5%) below 230℃

DOI: 10.1051/matecconf/20166402002

as represented in Fig. 11. So that the PBA was rarely broken down below 230℃. In addition, the yield of the chain extender BDO could reach to 93% (Fig. 12). To sum up, in subcritical methanol(230 ℃ /7.2MPa), the yields of MDC and BDO were high, and the decomposition rate of PBA was low.

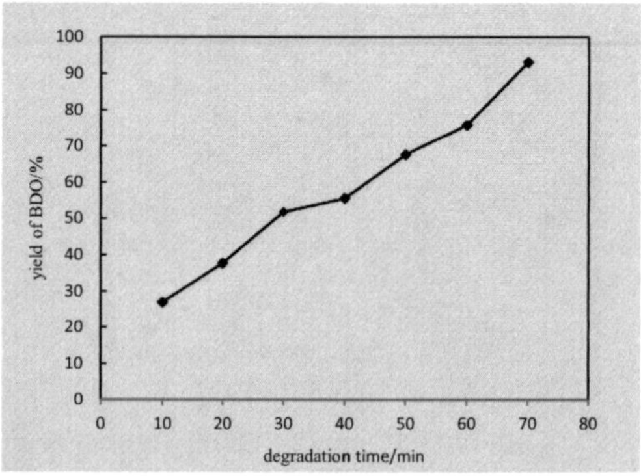

Figure 12. Relationship between yield of extender(BDO) and degradation time

3.2.3 Characterization of PBA

According to the study in 3.2.1 and 3.2.2, 230℃ /7.2MPa was the appropriate condition of TPU degradation in methanol. But one of the main products PBA was polymer, it needed to be measured by GPC.

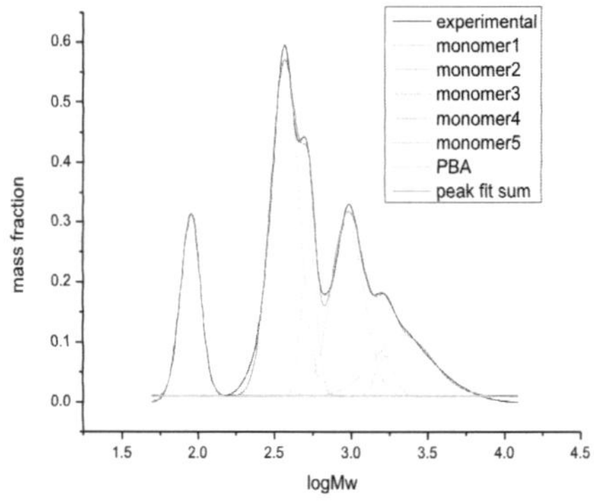

Figure 13. Molecular weight distribution of degradation products in 230℃/20min

The molecular weight distributions (MWDs) of the products mixture at 230℃ were analyzed by GPC. Take the MWD at 20min as an example, as represented in Fig. 13, the peak of PBA could be separated from the monomer peaks by origin8.5 PFM. The MWDs of PBA at different times were shown in Fig. 14. The molar concentrations of PBA were also obtained by calculation (Fig. 15). The maximum yield was 92%.

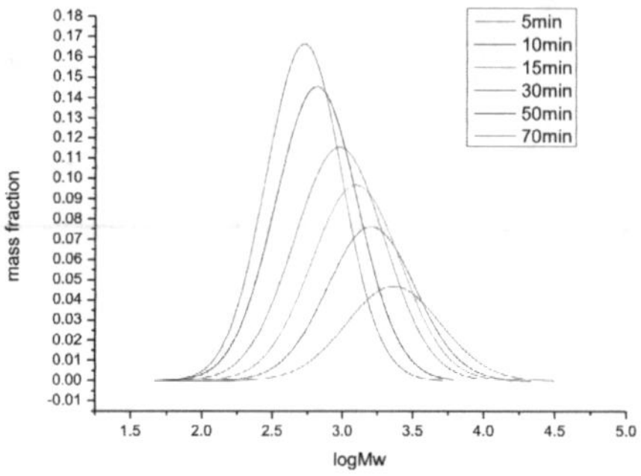

Figure 14. Relationship between MWD of PBA and degradation time

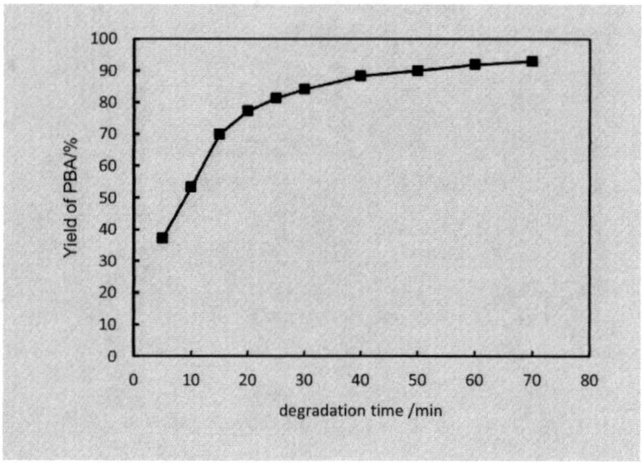

Figure 15. Relationship between yield of PBA and degradation time

4 Conclusions

Our research realized that supercritical methanol performed excellently in TPU degradation. TPU decomposed efficiently to obtain its raw material MDC, PBA and chain extender BDO. In the optimum degradation condition 230℃/7.2MPa, the yields of TPU MDC, PBA, BDO could reach 98%, 92% and 93%, respectively。

References

1. Günter Oertel. *Polyurethane Handbook* .Hanser .Publishers,(1985).
2. khalid mahmood zia, haq nawaz bhatti , ijaz ahmad bhatti. Methods for polyurethane and polyurethane composites, recycling and recovery: a review. React Funct Polym. **67**, 675–6923(2007).

3. chao-hsiung wu, ching-yuan chang, jiunn-kuen li. Glycolysis of rigid polyurethane from waste refrigerators. Polym Degrad Stabil. **75**,413–421(2002).
4. chao-hsiung wu, ching-yuan chang, chien-min cheng, hung-chang huang， glycolysis of waste flexible polyurethane foam. Polym Degrad Stabil. **80**,103–111(2003).
5. tsukahara t, harada m, tomiyasu h, ikeda y,j phy chem a, **112**, 9657-9664(2008).
6. lalanne p, andanson j m, soeten j c, *et al*. J Chem Phys A, **108**,3902-3909(2004).
7. M. Goto, H. Koyamoto, A. Kodama, T. Hirose, S.Nagaoka, B.J. McCoy, Degradation kinetics of poly(ethylene terephthalate) in supercritical methanol, AIChE Journal.**48**,136–141(2002).
8. M. Genta, T. Iwaya, M. Sasaki, M. Goto, Supercritical methanol for polyethylene terephthalate depolymerization: observation using simulator. Waste Manage 27,1167–1177(2007).
9. Lei Chen. *Depolymerization of Polyethylene terephthalate and Polycarbonate in Supercritical Methanol.* Shang hai: East China University of Science and Technology(2003).
10. Huang Jie，Huang Ke, Qi Wenjie, Zhu Zibin. Process analysis of depolymerization polybutylene terephthalate in supercritical methanol. Polym Degrad Stabil **91**, 2527-2531(2006).
11. Jie Huang，Junhe Yang, Minking K. Chyu, Qingming Wang, Zibin Zhu. Continuous-distribution kinetics for degradation of polybutylene terephthalate (PBT) in supercritical methanol. Polym Degrad Stabil **94**: 2142–2148(2009).
12. Lu Liu, Lihua Tang, Yongqiang Wu, Yanhui Ni, Zibin Zhu, Degradation process investigation of thermoplastic polyurethane elastomer in supercritical methanol. Polym Degrad Stabil.**98**,2520-2528(2013).

DOI: 10.1051/matecconf/20166402003

Water-Sensitivity Characteristics of Briquettes Made from High-Rank Coal

Yunguang Geng, Dazhen Tang, Hao Xu and Tang Shuling

School of Energy Resources, China University of Geosciences (Beijing), Beijing 100083, China

Abstract. In order to study the water sensitivity characteristics of the coalbed methane (CBM) reservoir in the southern Qinshui Basin, the scanning electron microscopy, mineral composition and the water sensitivity of main coalbed 3 cores were tested and analyzed. Because CBM reservoirs in this area are characterized by low porosity and low permeability, the common water sensitivity experiment of cores can't be used, instead, the briquettes were chose for the test to analysis the water sensitivity of CBM reservoirs. Results show that: the degree of water sensitivity in the study area varies from week to moderate. The controlling factors of water sensitivity are clay mineral content and the occurrence type of clay minerals, permeability and liquid flow rate. The water sensitivity damage rate is positively correlated with clay mineral content and liquid flow rate, and is negatively correlated with core permeability. The water sensitivity of CBM reservoir exist two damage mechanisms, including static permeability decline caused by clay mineral hydration dilatation and dynamic permeability decline caused by dispersion/migration of clay minerals.

1 Introduction

Coalbed methane (CBM) is a natural gas that gathers in pores and microfractures of coal seams [1]. A coalbed methane reservoir is one kind of unconventional petroleum resource where coalbed gas is developed and stored, unlike conventional natural gas reservoirs [2, 3]. Coalbeds are characterized by low mechanical strength, cleat development, strong heterogeneity and a large specific surface area compared with conventional sandstone rock. Coalbed permeability is the important factor affecting CBM well production, which is not only controlled by the factors such as coal lithotypes, pores and cleats characteristics, but also affected by different kinds of sensitivities [4-6].

Tectonic coal exists in the area with intense tectonic activity in Qinshui Basin. Based on the coal seam has the minimum strength in the coal-bearing rock series, the original structure of coal is easily destroy in the process of structural changes forming tectonic coal. According to the damage degree, tectonic coal contains fractured, granular and powdered, and characterized with disorder bedding, soft coal property, relatively high porosity and destruction or disappearance of the original structure. The structure and composition of briquette is familiar with tectonic coal, which can be applied to analysis the properties of granular tectonic coal. Huge advantage of the coal briquettes application is a possibility of obtaining a material of repeatable properties [7]. In addition the study presented in the current paper aims the coal briquetting as the important process for the outburst investigation and the process that is helpful in exploration of the dangerous coal structures which appear in the natural conditions

2 Experimental methods

2.1 Materials

Coal samples were collected from the NO.3 coal seam in Southern Qinshui Basin. Briquettes were made of artificially selected certain particle grain size of coal particles which ranges from 0.2~0.4 mm. Briquettes are 3~3.5 cm long and with diameter of 2.5 cm (Table 1). Making process is as follows:

(1) Break the coal samples and select 0.2~0.4 mm coal particles.

(2) Take 20g particles and add 2mL standard brine mixing with coal particles.

(3) Put the particles into mode and exert 17500N pressure on it for 6 minutes.

Table 1. Basic parameters of the briquette samples.

Samples	Size /cm		permeability /mD	porosity /%
	diameter	length		
1	2.5	3.241	11.89	9.2
2	2.5	3.432	17.32	11.4
3	2.5	3.501	14.88	9.8
4	2.5	3.156	14.31	10.1
5	2.5	3.327	21.63	12.9

2.2 Methods

2.2.1 Mineral, microstructure and pore diameter analysis of coal samples

The whole-rock mineral and relative clay mineral contents of the coal seam were determined by X-ray diffraction based on the Rietveld method [8, 9]. Microstructures of fresh sections were observed using an S-4800 scanning electron microscope (SEM) by secondary electron imaging [10].

2.2.2 Formatting author names and author affiliations

The experiment measured permeability under the flow rate of 1mL/min, 3mL/min, and 5mL/min respectively under the condition of saturated 8% KCl and distilled water. The permeability measurement references to Darcy formula. The differences of permeability measured under the same flow rate with the two conditions are compared. Water sensitivity damage rate is as the follow equation:

$$D_W = \frac{|K_i - K_w|}{K_i} \times 100\% \quad (1)$$

where Dw is sensitivity damage rate; Ki is permeability measured under distilled water condition, 10-3μ m2; Kw is permeability measured under 8% KCl condition, 10-3 μ m2.

The experiment process is as follows:

(1) Put briquette sample into 8% KCl for 48 hours to make it saturated.

(2) Put the saturated 8% KCl sample in LDY32-300 core flow instrument and increasing confining pressure to 3.5MPa.

(3) Conduct the displacement test using 8% KCl, measuring permeability under the flow rate of 1mL/min, 3mL/min, and 5mL/min respectively.

(4) Take out the sample and put it into distilled water for 48 hours to make it saturated.

(5) Repeat step (2) and step (3).

3 Results and discussion

3.1 Mineral composition and occurrence manner

The microstructure of the NO.3 coal seam in Qinshui Basin had the following characteristics: parallel grains, bedding and joints. The cleats (Fig. 1e) of the coal were easily recognized and showed good tropism with grid shaped distributions on the flat surface Micropores, including blow holes, moldic pore and deformed tissue holes, developed locally (Fig. 1a, b, d); and the pores were partially filled by clay minerals (Fig. 1d). A large number of microfractures existed (Fig. 1f), with moderate connectivity. The main fissures were 0.14-1.1 cm long and 270 mm wide. And several microfractures were also filled by clay minerals (Fig. 1f). This is the reason why coal seam characterized by low porosity and permeability. Clay minerals in micropores will hydrate once in contact with water, leading to flow path and particulate output reduction.

Vitrinite reflectance of samples was greater than 2.6. Maceral group composition contains vitrinite and collinite with no exinite existing. Non mineral composition is given priority to clay minerals and contained small amounts of carbonate (Table 2). Under the scanning electron microscopy (SEM), clay minerals such as kaolinite shaped in very thin flaky on the surface coal rock (Fig. 1c). Some epigenetic clay minerals filled in coal pores and fractures (Fig. 1d, f).

Table 2. Mineral compositions determined by X-ray diffraction.

Samples	Non minerals (%)	Clay (%)	Calcite and Dolomite (%)	Siderite (%)	Others (%)
1	96.4	2.3	0.5	0.4	0.4
2	94.3	5.3	0.2	0.2	/
3	91.7	7.1	0.3	0.9	/
4	95.1	4.3	0.2	0.5	/
5	88.9	10.3	0.2	/	0.6

3.2 Water-sensitivity of coal samples

Water-sensitivity test result showed that the permeability damage ranged from 29.92%~54.03%, the average of 43.53% (Fig. 2). The degree of water-sensitivity damage at 5 briquette samples varied from weak to moderate with no strong water-sensitivity appearance.

3.2.1 Effect factors of water-sensitivity

Water-sensitivity depends on the type and content of clay minerals in reservoir and underground fluid salinity. Predecessors attributed the reservoir water-sensitivity to the hydration swelling and dispersion of the clay minerals mingling under the condition of low salinity. In common clay minerals, montmorillonite has the strongest swelling capacity, followed by illite/montmorillonite mixed layer and chlorite/smectite mixed layer, because swelling ability of illite and chlorite is relatively weak, while kaolinite has no swelling ability.

The contents of clay and carbonate minerals and water-sensitivity damage rate correlation analysis result shows that the coal reservoir water sensitivity damage rate is positively correlated with clay mineral content, and is negatively related with carbonate content (Fig. 3). It suggests that the main factors causing the reservoir sensibility in this area are clay minerals. The more the pores and fractures are filled with clay mineral, the more the water-sensitivity damage rate will be.

Shown by analysis, water-sensitivity damage rate is positively correlated with flow rate (Fig. 4). Along with

DOI: 10.1051/matecconf/20166402003

the increase of the displacement velocity water-sensitivity damage rate increases, indicating that water-sensitivity minerals are not only cause the static permeability decline, but also have a certain influence to dynamic permeability decline.

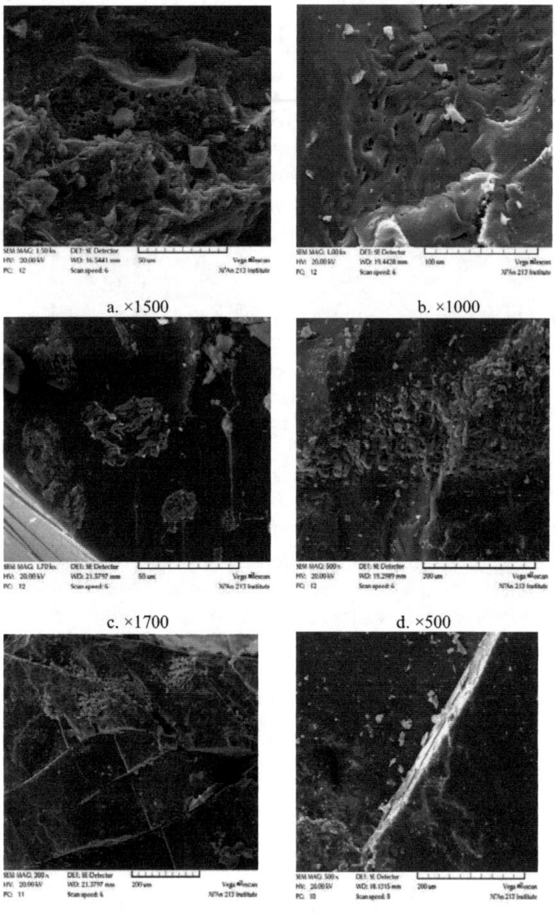

Figure 1. Microscopic images of the coal core of the NO. 3 coal seam.

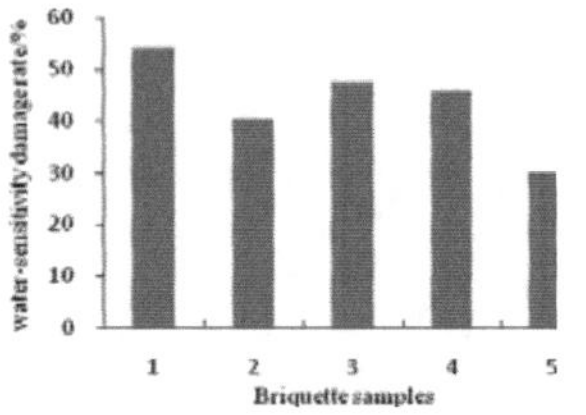

Figure 2. Water-sensitivity damage rate of briquette samples.

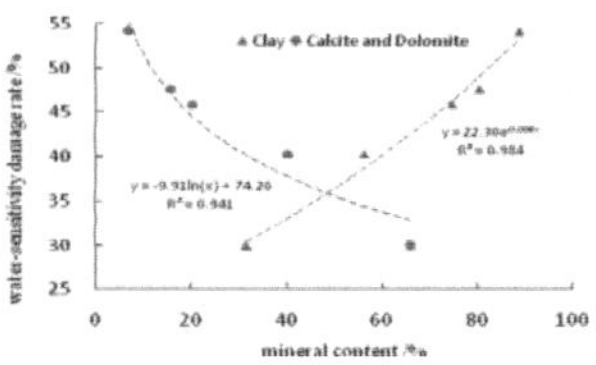

Figure 3. The relationship between water-sensitivity damage rate and mineral content under the same flow.

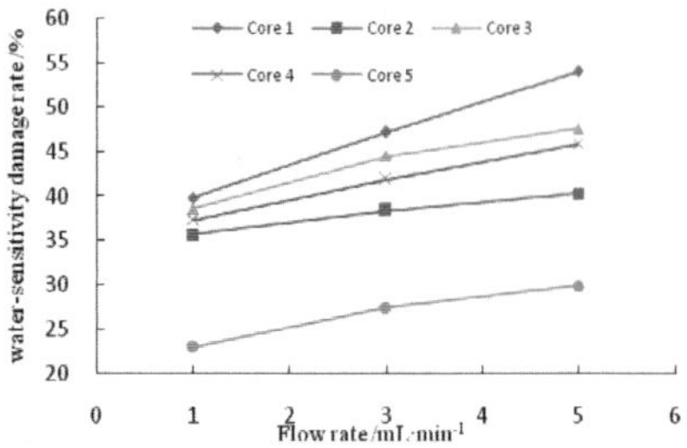

Figure 4. The relationship between water-sensitivity damage rate and flow rate.

4 Conclusions

(1) Maceral group composition of coal samples contains vitrinite and collinite with no exinite existing. Non mineral composition is given priority to clay minerals and contained small amounts of carbonate. Coalbed No.3 in Qinshui Basin is characterized by low porosity and low permeability; pore type is major in micropore; micro-fractures development, which mostly are secondary cracks with varying ages. The effective fracture length is short and with high closure degree and fillings of mineral, which cause the overall coalbed reservoir permeability is low.

(2) The coal rock from the NO.3 seam was mainly composed of carbon and a certain amount of clay minerals, making it sensitive to water. Briquettes made from high rank coal exists the following two water-sensitivity mechanisms. One is under low flow rate condition, clay mineral hydration expansion blocking pore throat leads to the static permeability decline. The other is under high flow rate condition, in addition to hydration expansion mechanism, coal particles migration leads to dynamic permeability decline.

(3) Flow experiments result shows a positive correlation between cores water-sensitivity damage rate and clay mineral content and a negative correlation between the damage rate and calcite and dolomite content under the same flow rate. In addition, there is a positive correlation between the water-sensitivity damage rate and flow rate.

References

1. O. Karacan, E. Okandan, Assessment of energetic heterogeneity of coals for gas adsorption and its effect on mixture predictions for coalbed methane studies, Fuel **79**, 1963-1974 (2000)
2. A. Keim, D. Luxbacher, M. Karmis, A numerical study on optimization of multilateral horizontal wellbore patterns for coalbed methane production in southern Shanxi province, China, Int. J. Coal Geol. **86**, 306-317 (2011)
3. H. Xu, D. Z. Tang, S. H. Tang, J. L. Zhao, et al., A dynamic prediction model for gas–water effective permeability based on coalbed methane production data, Int. J. Coal Geol. **121**, 44–52 (2014)
4. H. Kumar, D. Elsworth, J. S. Liu, et al., Optimizing enhanced coalbed methane recovery for unhindered

production and CO_2 injectivity, Int. J. Greenh. Gas. Control **11**, 86-97 (2012)

5. S. Rachmat, A. Pramana, L. Febriana, Indonesia's unconventional resources, modified resource triangle, and a typical example of stimulation of coalbed methane reservoir, Mod. Appl. Sci. **6**, 99-111 (2012)
6. Z. J. Pan, D. Connell, M. Camilleri, Laboratory characterisation of coal reservoir permeability for primary and enhanced coalbed methane recovery. Int. J. Coal Geol. **82**, 252-261 (2010)
7. B. Dutka, M. Kudasik, Z. Pokryszka, N. Skoczylas, J. Topolnicki, M. Wierzbicki, Balance of CO2/CH4 exchange sorption in a coal briquette, Fuel Process Technol. **106**, 95-101 (2013)
8. C. R. Ward, J. C. Taylor, C. E. Matulis, et al., Quantification of mineral matter in the Argonne premium coals using interactive Rietveld-based X-ray diffraction, Int. J. Coal Geol. **46**, 67-82 (2001)
9. C. D. Ruan, C. R. Ward, Quantitative X-ray powder diffraction analysis of clay minerals in Australian coals using Rietveld methods, Appl. Clay Sci. **21**, 227-240 (2002)
10. Y. D. Cai, D. M. Liu, Y. B. Yao, et al., Partial coal pyrolysis and its implication to enhance coalbed methane recovery, part I: an experimental investigation, Fuel **132**, 12-19 (2014)

Molecular cloning and expression analysis of mevalonate pyrophosphate decarboxylase in *Antrodia cinnamomea*

Jing Li[1,3], Xiong-Jie Lin[2], En-Si Shao[1,3] and Zhan-Xi Lin[1,3]

[1]*Fujian Agriculture and Forestry University, Fuzhou 350002, China*

[2]*Fruit Research Institute, Fujian Academy of Agriculture Sciences, Fuzhou 350013, China*

[3]*China National Engineering Research Center of JUNCAO Technology, Fuzhou 350002, China*

Abstract. Mevalonate pyrophosphate decarboxylase (MVD) plays important roles in triterpenoid biosynthesis via the mevalonate pathway. A novel MVD gene was isolated and identified in *Antrodia cinnamomea* (Ac-mvd). The full-length cDNA contained an open reading frame with a length of 1,209 bp and encoded a 402-amino acid polypeptide with a molecular mass of 43.3 kDa and a theoretical pI of 8.23. As incubation time was prolonged for 28 days, the triterpenoid content in the mycelium gradually increased and reached 39.192 ± 2.025 mg/g; the triterpenoid content of the fruiting body grew in the hay of *Cinnamomum kanehirae* (ACFB-CK), was 49.391 ± 2.675 mg/g, which was significantly different from other samples ($P < 0.05$). qRT-PCR revealed that the highest expression levels of Ac-mvd in the mycelium of *Antrodia cinnamomea* were detected on the 7th day. The expression levels gradually decreased as culture time was extended from 14 days to 42 days.

1 Introduction

Antrodia cinnamomea (*A. cinnamomea*) is a traditional medicinal fungus in Taiwan and in the surrounding areas in Asia; this fungus inhabits the empty brown rot cavity of *Cinnamomum kanehirae* [1]. *A. cinnamomea* is expensive because of several characteristics, such as high medicinal value, scarcity in nature, and difficulty in artificial cultivation. The fruiting body and mycelium extracts of *A. cinnamomea* have been used as a medicine to improve the activity of organisms in studies involving cell lines and animal models *in vivo* and *in vitro*; *A. cinnamomea* extracts provide anti-oxidative [2], anticancer, anti-hepatitic [3], and anti-inflammatory properties [4]. Previous studies showed that the main active ingredients of *A. cinnamomea* are antroquinonol [5], succinic acid, maleic acid [6], benzenoids [7], triterpenoids [8], and polysaccharides [9]. Among these components, triterpenoid constitutes the largest group of phytochemicals in all living organisms [10]; as such, studies have mainly focused on this group. Most triterpenoids are isolated from plants and utilized as commercial flavorings and medicines. Martin et al. [11] devised an alternative method to produce triterpenoids by using *Escherichia coli*.

Previous studies on plants and fungi confirmed that triterpenoids are synthesized via the mevalonate (MVA) pathway [12]. In 2014, the triterpenoid biosynthesis pathway in *A. cinnamomea* was validated through genomic and transcriptome analyses; this pathway is similar to the MVA pathway in *Ganoderma lucidum* [13]. Several enzymes, such as squalene synthase (SQS) [14], HMG-CoA reductase (HMGR) [15], mevalonate pyrophosphate decarboxylase (MVD) [16], and farnesyl-diphosphate synthase (FPS) [17], are implicated in the MVA pathway. A previous study showed that the synthesis of triterpenoids in *A. cinnamomea* is complex. MVD (EC 4.1.1.33) catalyzes the synthesis of mevalonate-5PP to form isopentenyl-PP, which accumulates to generate monoterpene [18]. Monoterpene is further converted by coenzyme Q (CoQ) to antroquinonol and ubiquinone; furthermore, SQS catalyzes the conversion of isopentenyl-PP to lanosterol. Lanosterol is an important substrate in the synthesis of ergot sterane and lanosterol triterpenoids by synthesizing cytochrome P450 [13]. The MVD activity has been quantified in animals [19], plants [20], and yeast. [21] MVD can be used as a potential target of several diseases, such as cancer and restenosis [22].

Few studies have focused on the characterization and expression of MVD in *A. cinnamomea* (*Ac-mvd*). In this study, *Ac-mvd* was cloned and characterized. This study aimed to investigate the expression profiles of *Ac-mvd* and the triterpenoid contents in various developmental stages of *A. cinnamomea* mycelium. The results can provide a basis of further studies on the gene function of the triterpenoid biosynthesis in *A. cinnamomea* and on the development of efficient fermentation processes involving triterpenoids.

DOI: 10.1051/matecconf/20166403001

2 Materials and methods

2.1 Strain and culture conditions

The *A. cinnamomea* AC001 (GenBank Accession No.: *KM925002*) was provided by Taiwan Shen-nong Fungal Co., Ltd. The fruiting body of *A. cinnamomea* grew in the hay of *C. kanehirae* (ACFB-CK). The fruiting body that grew in a corn cob (ACFB-CC) was stored in China National Engineering Research Center of JUNCAO Technology. For the shake-flask fermentation with GPM (2.5% glucose, 0.3% yeast extract, 0.5% peptone, 0.3% maltose, 0.1% KH_2PO_4, 0.1% $MgSO_4 \cdot 7H_2O$, 0.1% Vitamin B_1), 100 mL of the medium was placed in a 250 mL flask and then inoculated with 3% second-stage pre-cultured broth at 28 °C.

2.2 Nucleic acid extraction and cDNA synthesis

The mycelia of *A. cinnamomea* in GMP were harvested after 15 days, frozen in liquid nitrogen, and ground into powder in a mortar. Genomic DNA was extracted by using cetyltrimethyl ammonium bromide (CTAB) method previously described [23]. Total RNA was extracted from different *A. cinnamomea* mycelia and fruiting bodies by using TRIzol reagent (Invitrogen, California, USA). The quality and concentration of the samples were examined using NanoDrop 2000 (Thermo, New York, USA) and 1% agarose gel electrophoresis. cDNA was synthesized using a first-strand cDNA synthesis kit (TOYOBO, Tokyo, Japan). Each 10 μL of reaction mixture contained 500 ng of RNA, 2 μL of 4× DNase Master Mix, and RNase free water at 37 °C for 5 min; afterward, 2 μL of 5× RT Master Mix was mixed, and the appropriate buffer was added. The reaction mixture was incubated at 37 °C for 15 min, at 50 °C for 5 min, and at 98 °C for 5 min; then, the reaction mixture was subsequently stored at 4 °C.

2.3 Cloning of the specific fragment of *Ac-mvd*

On the basis of the conserved domains in the amino acid sequences of *Ganoderma. lucidum*, *Coriolus versicolor*, and *Pleurotus ostreatus*, we synthesized two degenerate oligonucleotide primers, namely, mvd-1 and mvd-2 (Table 1), for the *Ac-mvd* amplification of *A. cinnamomea* mycelial cDNA. The PCR amplification conditions were as follows: 95 °C for 5 min, followed by 32 cycles of denaturation at 94 °C for 30 s, annealing at 56.5 °C for 40 s, and extension at 72 °C for 7 min. The amplification products were transformed into a vector pMD™18-T (TaKaRa, Tokyo, Japan) and sequenced by Shanghai Biosune Biotechnology Co., Ltd.

2.4 Cloning of the full-length *Ac-mvd* through RACE

The sequence was used to design the primers for cloning the full-length sequence through the rapid amplification of cDNA ends (RACE).

Table 1. Oligonucleotide primers used in the experiment

Primer	5′–3′ Sequence
mvd-1	CGTGAACATCGCCTGCATNAARTAYTGGG (F)
mvd-2	CCCTGGACACGTCGTTCADRTARAADAT (R)
mvd-3	CGACGCCATTCAACCATAG
mvd-4	CCTTCTGGAATGACGGATC
mvd-5	GTTTCCGACGACAAGAAGGG (F)
mvd-6	TCCAGCACCGTATCGCACAT (R)
mvd-7	GCCCTCATTTCTTCATCC (F)
mvd-8	ATACACTCTAGCTCGCCATAA (R)
mvd-9	ACCCTTGACCAAGATCACTTTC (F)
mvd-10	GATATCTTCCTCGACGCCATTC (R)

Table 1 shows the PCR amplification conditions using the 5′ RACE protocol with Rapid Amplification of cDNA Ends (Invitrogen, California, USA) and 2 specific primers (mvd-3, and mvd-4). The mvd-4 primer was used for the self-formed adaptor (SEFA-PCR) [24], and mvd-4 was used for the nested PCR. The products were transformed into a pMD™18-T vector and sequenced.

The 3′ RACE sequences were amplified using the 3′-full RACE core set with primescript RTase kit and LA *Taq* with GC buffer (Takara, Tokyo, Japan). Two specific primers (mvd-5 and mvd-6; Table 1) were designed on the basis of the fragment sequence in accordance with the manufacturer's instructions. PCR amplification was carried out as follows: 94 °C for 2 min, 30 cycles at 98 °C for 10 s, 55 °C for 30 s, 72 °C for 2 min, and extension at 75 °C for 5 min. The product was purified, cloned into pMD™18-T, and sequenced.

The 3′ and 5′ RACE and fragment products of *A. cinnamomea* mycelia were aligned and assembled. The full-length sequence of *Ac-mvd* was amplified using specific primers mvd-7 and mvd-8 (Table 1) with gDNA and cDNA as template. The PCR amplification procedure was carried out as follows: 94 °C for 5 min, followed by 32 cycles of denaturation at 94 °C for 30 s, annealing at 52.8 °C for 30 s, and elongation at 72 °C for 40 s, with extension at 72 °C for 7 min. The product was purified, cloned into pMD™18-T, and sequenced.

2.5 Bioinformatics analysis

The full-length sequence of *Ac-mvd* was identified, and homologs were analyzed in accordance with the BLASTP algorithm of NCBI (http://blast.ncbi.nlm.nih.gov/Blast.cgi). Multiple alignments were carried out using Clustal X 2.0 and the MEGA 5.1 [25]. Data were analyzed using the Maximum Likelihood Method (ML) [26], and the bootstrap strength was supported by 1000 replicates for both in ML analysis. The complete amino acid sequences of each fungus were used to construct the phylogenic tree. The protein sequences were analyzed using the bioinformatics tools ExPASy, SignalP4.1 Sever, and TMHMM Server2.0. The secondary and tertiary structures of the *Ac-mvd* protein were predicted using the SWISS-MODEL Protein Modeling Server (http://swissmodel.expasy.org/) and InterPro Protein (http://www.ebi.ac.uk/interpro/). Further analysis was carried out using PyMOL.

2.6 Expression analysis of *Ac-mvd* through quantitative real-time PCR (qRT-PCR) in different development stages

The mycelia of *A. cinnamomea* were harvested after the fungus was cultured in GPM for 7, 14, 21, 28, 35, and 42 days. Total RNA was isolated from each stage, and the synthesized cDNA was used as a template for qRT-PCR amplification with two primers, namely, mvd-9 and mvd-10 (Table 1). qRT-PCR was performed using CFX Connect (Bio-rad, California, USA) in 20 μL of reaction mixture containing 10 μL of SYBR Premix Ex Taq™ (TaKaRa, Tokyo, Japan), 1 μL of sample cDNA, 2 μL of primers, and 7 μL of sterile double-distilled water. The cDNA of ACFB-CK was used as a positive control. Double-distilled water instead of first-strand cDNA was used as a negative control.

2.7 Extraction and analysis of triterpenoids in the mycelium and fruiting body of *A. cinnamomea*

The total triterpenoids of *A. cinnamomea* mycelium were extracted through ultrasonic treatment (Kun Shan Ultrasonic Instruments Co., Ltd., Jiangsu, China). Dried mycelia and fruiting body powder (0.5 g) were mixed with 40 mL of 80% ethanol and ultrasonic wave at 70 °C for 2 h. The slurry was centrifuged at 7000 rpm for 20 min to collect the supernatant, and the supernatant volume was diluted with 80% ethanol to 50 mL.

Each sample (0.5 mL) was placed in a scale tube and dried at 50 °C. Afterward, 0.3 mL of 5% vanillic-acetic acid and 1 mL of perchloric acid were added to the tubes, and the reaction was mixed at 60 °C for 20 min. Acetic acid (10 mL) was added to the tubes, and the reaction was mixed after cooling [27]. The total triterpenoid content was determined by measuring the absorbance at 548 nm with a spectrophotometer (Biochrom, Cambridge, England), and oleanolic acid was used as the standard. The concentration of triperpenoids was calculated using Eq. (1):

$$Total\quad triperpendiod(mg/g) = \frac{triperpendiodsqualityinsample}{\frac{weightofsample}{constantvolume} \times volumeofsample} \tag{1}$$

2.8 Statistical analysis

Data were expressed as the average of three independent sample measurements. Error bars indicated the standard deviations from the triplicates. The difference between contrasting treatments was considered significant at $P < 0.05$. *Ac-mvd* levels were determined according to a standard equation curve ($Y = -3.1854X + 43.264$, $R^2 = 0.996$, Log Starting quantity (Log SQ) is X, Cq Value is Y) [28]. The gene copies were calculated using Eq. (2).

$$Gene\ copies\ (copies/\mu L) = 6.02X10^{23} \square (500\ ng/\mu L)/cDNA\ Length \square 660 \tag{2}$$

3 Results

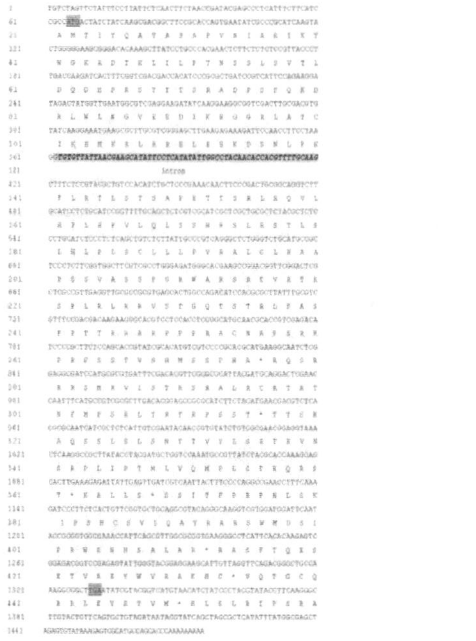

Figure 1. Promoter region, genomic DNA, and predicted amino acid sequence of *Ac-mvd*. The intron of mvd was shaded in bold italics and light gray.

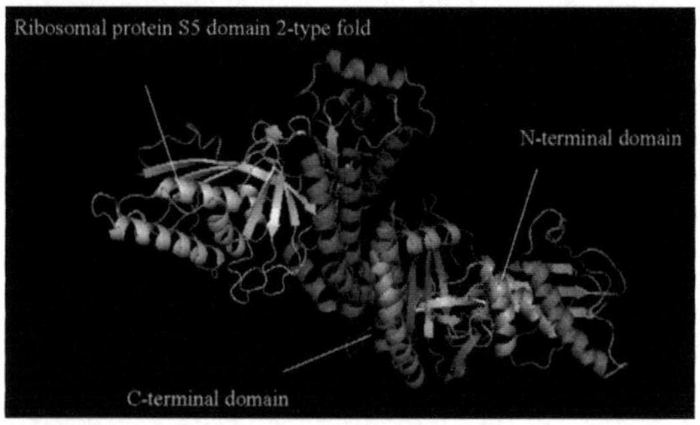

Figure 2. 3-D structural model of the amino acid sequences of Ac-mvd. The structural model of *Ac-mvd* was created on the basis of *Gl-mvd* (3f0n.1.A). The pI of MVD is 8.23, and no signal peptide and transmembrane domain are detected. The hydrophobic and hydrophilic regions are shown in the protein conformation.

3.1 Cloning and Characterization of *Ac-mvd*

A novel full-length cDNA clone of *Ac-mvd* was generated from *A. cinnamomea*. The 809 bp fragment of *Ac-mvd* was obtained from cDNA via PCR by using two degenerate primers, namely, mvd-1 and mvd-2. The product was homologous to *mvd* in *G. lucidum*, *Trametes cinnabarina*, *Trametes versicolor*, *Ceriposiospis subvermispora*, and *Dichomitus squalens*. The 3′ and 5′ RACE products were amplified depending on the obtained specific 809 bp fragment. The product length of 3′ RACE was approximately 690 bp; the product length

DOI: 10.1051/matecconf/20166403001

of 5′ RACE was approximately 197 bp. The full-length *Ac-mvd* was obtained using the mvd-7 and mvd-8primers with cDNA and gDNA as the respective templates. The 1268 bp full-length gDNA was composed of a 59 bp intron (Fig. 1). The full-length cDNA encoded a 402-amino acid polypeptide with a theoretical molecular mass of 43.3 kDa. The polypeptide contained the distinctive domain of the ribosomal protein S5 domain; the polypeptide was also similar to the MVD of *G. lucidum* (Fig. 2). The nucleotide sequence of cDNA (*Ac-mvd*) was deposited in GenBank with the accession number *KR364808*.

3.2 Phylogenetic analysis

A phylogenetic tree was generated on the basis of the deduced amino acid sequences to determine the evolutionary relationship between *Ac-mvd* and other fungi; the BLASTX search showed that the *Ac-mvd* fragment was highly homologous to the other fungi (Fig. 3). Most of the predicted homologous proteins of *Ac-mvd* in *D. squalens* (88% identity), *G. lucidum* (86% identity), *T. versicolor* (85% identity), *Rhizoctonia solani* (79% identity), and *Cryptococcus gattii* (70% identity) were selected and subjected to phylogenetic analysis. *mvd* is generally found in many fungi, whereas the homologous *mvd* sequences were divided into the following groups: basidiomycetes (*G. lucidum* and *T. versicolor*) and yeast (*Schizosaccharomyces cryophilus* and *C. gattii*).

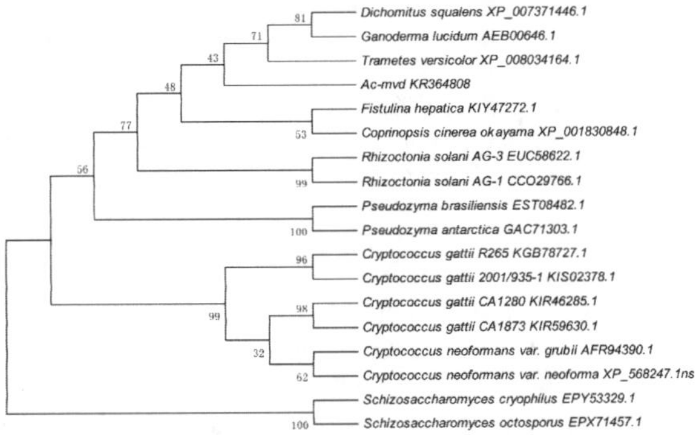

Figure 3. Phylogenetic tree showing the relationship of *Ac-mvd* with other proteins from various fungi.

3.3 Ac-mvd expression and triterpenoid concentration in the mycelium of *A. cinnamomea* in different development stages

The cDNA was synthesized as a template to determine the expression levels of *Ac-mvd* in different stages through absolute qRT-PCR. The gene expression levels of *Ac-mvd* were calculated from 7 days to 42 days; the results revealed that the expression levels decreased sharply from 47.3 × 10^5 copies/μL to 35.5 × 10^4 copies/μL. ACFB-CK and ACFB-CC contained 49.8 × 10^4 and 21.1 × 10^4 copies/μL *Ac-mvd*, respectively (Fig. 4).

The submerged culture of *A. cinnamomea* mycelia and fruiting bodies was dried and used to extract and determine the triterpenoid content of the fungus. The triterpenoid contents in *A. cinnamomea* mycelia were 15.170 ± 1.400, 17.574 ± 2.015, 26.400 ± 2.259, 39.192 ± 2.025, 28.907 ± 0.619, and 30.074 ± 0.684 mg/g corresponding to the culture after 7, 14, 21, 28, 35, and 42 days (Fig. 5), respectively. As culture time was prolonged, the triterpenoid content of *A. cinnamomea* mycelia increased from 7 days to 28 days and then decreased from 35 days to 42 days. The triterpenoid contents of ACFB-CK and ACFB-CC were 49.391 ± 2.675 and 11.530 ± 0.733 mg/g, respectively, which were significantly higher than the triterpenoid contents in all of the development stages of the mycelia ($P < 0.05$) of other samples in ACFB-CK. By contrast, the lowest triterpenoid content was determined in ACFB-CC among the mycelium samples (Fig. 5).

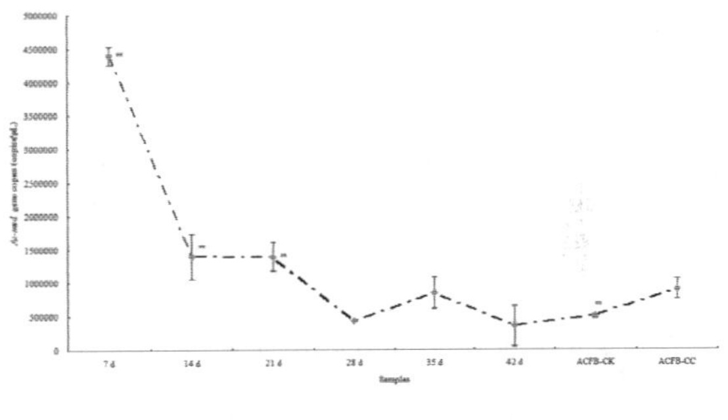

Figure 4. *A. cinnamomea* mycelia *Ac-mvd* gene copy numbers in different developmental stages (**$P < 0.05$)

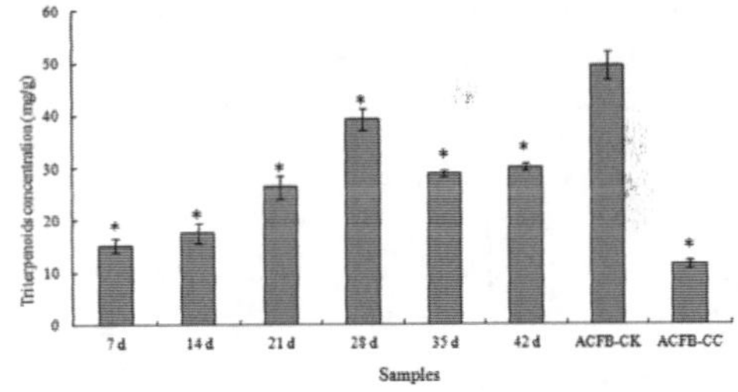

Figure 5. Triterpenoid contents of *A. cinnamomea* mycelia in a submerged medium in different stages; ACFB-CK was used as a control (* $P < 0.05$)

4 Discussion

Previous studies demonstrated that the triterpenoid biosynthesis of *A. cinnamomea* is significantly affected by several environmental factors, such as light, elicitor, substrate, pH value, and developmental stages [29-31]. The gene expression levels of *Ac-mvd* were the highest in the mycelia of *A. cinnamomea* on the 7th day of the submerged culture. The highest triterpenoid content was detected in ACFB-CK; conversely, the triterpenoid content of *A. cinnamomea* mycelia was slightly lower than that of ACFB-CK on the 28th day in the submerged culture. The lowest gene expression levels and triterpenoid content were detected in ACFB-CC. However, the absolute qRT-PCR results showed that the

Ac-mvd expression level was not linearly correlated with the triterpenoid content.

Triterpenoids exhibit important medicinal properties *in vivo* and *in vitro*. In a previous study, the overexpression of *mvd* is positively correlated with gene expression and triterpenoid content in *G. Lucidum*[16]. In the triterpenoid biosynthesis pathway of *A. cinnamomea*, MVD catalyzes the synthesis of mevalonate-5PP to form isopentenyl-PP, which is the substrate of zeatin and monoterpenoid biosynthesis. The main Kyoto Encyclopedia of Genes and Genomes (KEGG) mapping of the MVA pathway in *A. cinnamomea* shows that *Ac-mvd* is a differentially expressed gene in mycelia and fruiting bodies. To our knowledge, the *Ac-mvd* gene expression was significantly different between ACFB-CC and ACFB-CK in the mycelia cultured for 7 days. The triterpenoid content was decreased as culture time was prolonged. However, *Ac-mvd* expression was not directly proportional to *Ac-mvd* and triterpenoid, and this finding is distinct from that observed in *G. lucidum*. The MVA pathway is mainly involved in triterpenoid synthesis; the MVA pathway also provides a comprehensive collection of enzymes in the production of major metabolites. Metabolites can block polyisoprenoid and sterol biosynthesis in the MVA pathway [32]. Using 6-fluoromevalonic acid in tissue extracts, researchers found that MVD is the target of a downstream metabolite that blocks sterol production[33]. In our study, the triterpenoid content accumulated in the mycelium and *Ac-mvd* gene expression was reduced when the incubation time was extended.

In conclusion, a novel gene was cloned and characterized from *A. cinnamomea*. The *Ac-mvd* gene expression patterns in different stages of *A. cinnamomea* mycelia and fruiting bodies were also different. The triterpenoid content may be related to the gene expression level and accumulation time of the synthesis pathway. The successful cloning and expression analysis of *Ac-mvd* provide a framework for further studies on the function and triterpenoid synthesis in *A. cinnamomea*.

Acknowledgments

This work was supported by the Collaborative Innovation Center for Juncao Ecology Industry under Grant Fujian 2011 Program, k80nd8002 and the China National Engineering Research Center of JUNCAO Technology under Grant 2011FU125X12.

References

1. T.T. Chang, W.N. Chou. Mycological Research, 1995, **99**(6).
2. C.H. Shu, M.Y. Lung. Journal of the Chinese Institute of Chemical Engineers, 2008, **39**(1).
3. Y.C. Ho, M.T. Lin, K.J. Duan, Y.S. Chen. Journal of the Chinese Institute of Chemical Engineers, 2008, **39**(5).
4. C.L. Wen, C.C. Chang, S.S. Huang, C.L. Kuo, S.L. Hsu, J.S. Deng, G.J. Huang. Journal of ethnopharmacology, 2011, **137**(1).
5. V.B. Kumar, T.C. Yuan, J.W. Liou, C.J. Yang, P.J. Sung, C.F. Weng. Mutation Research/Fundamental and Molecular Mechanisms of Mutagenesis, 2011, **707**(1).
6. W.L. Lin, Y.J. Lee, S.M. Wang, P.Y. Huang, T.H. Tseng. European journal of pharmacology, 2012, **680**(1).
7. H.M. Lien, H.W. Lin, Y.J. Wang, L.C. Chen, D.Y. Yang, Y.Y. Lai, Y.S. Ho. Evidence-Based Complementary and Alternative Medicine, 2011, **2011**.
8. T.W. Ma, Y. Lai, F.C. Yang. Bioprocess Biosyst Eng, 2014.
9. M.C. Tsai, T.Y. Song, P.H. Shih, G.C. Yen. Food chemistry, 2007, **104**(3).
10. A. Bishayee, S. Ahmed, N. Brankov, M. Perloff. Frontiers in bioscience: a journal and virtual library, 2011, **16**.
11. V.J. Martin, D.J. Pitera, S.T. Withers, J.D. Newman, J.D. Keasling. Nature biotechnology, 2003, **21**(7).
12. T. Goodwin. Biochemical Journal, 1971, **123**(3).
13. M.Y.J. Lu, W.L. Fan, W.F. Wang, T.C. Chen, Y.C. Tang, F.H. Chu, T.T. Chang, S.Y. Wang, M.Y. Li, Y.H. Chen. Proceedings of the National Academy of Sciences, 2014, **111**(44).
14. J.S. Zhou, S.L. Ji, M.F. Ren, Y.L. He, X.R. Jing, J.W. Xu. Biochemical Engineering Journal, 2014, **90**.
15. C.H. Shang, F. Zhu, N. Li, X. Ou-Yang, L. Shi, M.W. Zhao, Y.X. Li. Bioscience, Biotechnology, and Biochemistry, 2008, **72**(5).
16. L. Shi, L. Qin, Y. Xu, A. Ren, X. Fang, D. Mu, Q. Tan, M. Zhao. Molecular biology reports, 2012, **39**(5).
17. Y.X. Ding, X. Ou-Yang, C.H. Shang, A. Ren, L. Shi, Y.X. Li, M.W. Zhao. Bioscience, Biotechnology, and Biochemistry, 2008, **72**(6).
18. K. Bloch, S. Chaykin, A. Phillips, A. De Waard. Journal of Biological Chemistry, 1959, **234**(10).
19. M. Alvear, A.M. Jabalquinto, J. Eyzaguirre, E. Cardemil. Biochemistry, 1982, **21**(19).
20. D. Skilleter, R. Kekwick. Biochem. J, 1971, **124**.
21. J.K. Bartz, D. Söll. Biochimie, 1972, **54**(1).
22. M.J. Toth, L. Huwyler. Journal of Biological Chemistry, 1996, **271**(14).
23. M.A. Saghai-Maroof, K.M. Soliman, R.A. Jorgensen, R.W. Allard. Proceedings of the National Academy of Sciences of the United States of America, 1984, **81**(24).
24. S. Wang, J. He, Z. Cui, S. Li. Appl Environ Microbiol, 2007, **73**(15).
25. K. Tamura, D. Peterson, N. Peterson, G. Stecher, M. Nei, S. Kumar. Molecular biology and evolution, 2011, **28**(10).
26. A. Jagerschmidt, N. Guillaume, B.P. Roques, F. Noble. Molecular pharmacology, 1998, **53**(5).
27. Z.M. Lu, W.Y. Tao, H.Y. Xu, J. Lim, X.M. Zhang, L.P. Wang, J.H. Chen, Z.H. Xu. Food chemistry, 2011, **127**(2).
28. P. Cawthon's quantitative real-time. Biotechniques, 2008, **44**.
29. F.C. Yang, H.C. Huang, M.J. Yang. Enzyme and Microbial Technology, 2003, **33**(4).

30. C.H. Shu, M.Y. Lung. Journal of the Chinese Institute of Chemical Engineers, 2008, **39**(1).
31. C.J. Liu, C.C. Chiang, B.H. Chiang. Biochemical Engineering Journal, 2012, **64**(0).
32. H.M. Miziorko. Archives of biochemistry and biophysics, 2011, **505**(2).
33. J. Nave, H. d'Orchymont, J. Ducep, F. Piriou, M. Jung. Biochem. J, 1985, **227**.

 DOI: 10.1051/matecconf/20166404001

Conductivity of Cellulose Acetate Membranes from Pandan Duri Leaves (*Pandanus tectorius*) for Li-ion Battery

Endang W. Laksono, Marfuatun and Demas Aji

Department of Chemistry, Faculty of Mathematics and Natural Sciences, Yogyakarta State University, Yogyakarta 55281, Indonesia

Abstract. The purpose of this research is to know the influence of lithium chloride composition on membrane conductivity. Cellulose was extracted from pandan duri leaves *(P. tectorius)* by dilute alkaline and bleaching with 0.5% NaOCl followed by synthesis of cellulose acetate using acetic anhydride as acetylating agent, acetic acid as solvent and sulfuric acid as catalyst. The membranes were prepared by casting polymer solution method and the composition of CA/LiCl were 60/40, 65/35, 70/30, 75/25, 80/20 and 100/0. Structural analysis was carried out by FTIR and X-ray diffraction. The conductivity was measured using Elkahfi 100. The highest conductivity of cellulose acetate membrane was 2.20 x 10^{-4} S cm^{-1} that measured at room temperature for 65/35 composition

1 Introduction

Nowadays, many people in this century are using portable electronic devices such as smartphone, smartwatch, digital camera, laptop and tablet to support their daily activities. Many of them are powered by lithium ion (Li-ion) battery due to its high energy density, lack of memory effect, long cycle life, high charge and discharge rate capabilities [1, 2]. Although it has many advantages, Li-ion battery has safety problem due to the using of liquid electrolyte that flammable and combustible, replacement of liquid electrolytes with polymer electrolytes is the best solution to the safety problem of li-ion battery [3].

Cellulose acetate (CA) membrane is the most common polymer electrolyte used, easily synthesized from cellulose of natural fibre [4]. The chemical method to obtain cellulose acetate is from esterification reaction of cellulose using acetic anhydride as an acetylating agent, acetic acid as a solvent and catalyst [5]. Polymer electrolytes like cellulose acetate has low conductivity, less than 10^{-6} S cm^{-1} [6, 7]. To enhance the conductivity of cellulose acetate, we can add 'salt' such as lithium chloride. The 'salt' is promising to increase the conductivity up to 10^{-3} S cm^{-1} due to the existence of lithium ion in the membrane [8].

In Indonesia, there are a lot of natural fibre plant. One of the natural fibre plant is Pandan duri. Pandan duri (*Pandanus tectorius*) is a plant from *Pandanaceae* family, which productively grow in this land. It called pandan duri because it has spiny margins on its leaves and 'duri' is refer to spine. Many people use the leaves as material for hand craft such as weaving (hat, bag, and mat) [9]. Pandan duri's leaves also has a potential to be develop as a raw material for cellulose acetate. According to Rasha M. Sheltami and her co-workers [10], pandan duri leaves contain a large amount of cellulose. Approximately 37.3% cellulose from wet pandan duri leaves and 81.6% after bleached [10]. The aim of this work is to know the influence of lithium chloride composition on membrane conductivity.

2 Experimental

2.1 Materials

Pandan duri leaves gathered in Bantul (Yogyakarta, Indonesia) were used in this research. The reagents and chemicals using in this research were standard analytical grades such as acetic acid, acetic anhydride, sodium hydroxide, lithium chloride, acetone, ethanol, hydrochloric acid and other chemicals. All chemicals were purchased from Chemix Laboratory and used without further purification.

2.2 Preparation of cellulose acetate membrane

First, cut the leaves into small pieces, removed the unused pieces (midribs and spiny margins), dried under the sun and ground. Next, the dry leaves were placed in stagnant water for three days. Then the leaves were boiled for about 20 min. The leaves were dried under the sun and ground again. Cellulose was isolated by soaking NaOH at room temperature and 80°C for 2 h. Next, bleached using 0.5% NaOCl. Then washed with distilled water for several times. Cellulose acetate was obtained by dissolved cellulose from pandan duri leaves in acetic acid and acetic anhydride with ratio of acetic acid to acetic anhydride was 70:27 (v/v). The ratio of cellulose to liquor was 5:112 (g/mL). Sulfuric acid was used as catalyst. Cellulose acetate membrane was prepared by dissolving cellulose acetate and LiCl in acetic acid with ratio 1:20 (g/mL). The composition of CA/LiCl were 100/0 (blank), 80/20, 75/25, 70/30, 65/35, and 60/40. Then, a mixture of CA and LiCl was casting on Teflon™ medium and dried at room temperature.

2.3 Determination of the degree of substitution (DS)

A mixture composed by 0.1 g cellulose from pandan duri leaves and 4 mL of 75% ethanol was heated at 60°C for 30 min. Then, 5 mL of 0.5 N NaOH was added to the mixture, which was heated at 60°C for 15 min. Next, the mixture was left to stand at room temperature for three days. After this time, the sample was titrated with 0.5 N HCl using phenolphthalein as the indicator and left to stand for 24 h. An excess of HCl was then back titrated with 0.5 N NaOH. A blank has same procedure determination for a control system of sample. The percentage of acetyl was calculated from Eq. (1):

$$Acetyl(\%) = [(V_d - V_c)N_{\text{NaOH}} + (V_a - V_b)N_{\text{HCl}}]x\frac{4.305}{w} \quad (1)$$

where, Va = NaOH volume required for titration of the sample; Vb = NaOH volume required for titration of the blank; Vc = HCl volume required for titration of the sample; Vd = HCl volume required for titration of the blank; w = weight of the sample. The degree of substitution (DS) of the sample was calculated from Eq. (2):

$$DS = \frac{3.86xAcetyl(\%)}{102.40 - Acetyl(\%)} \quad (2)$$

2.4 Determination of the viscosity-average molecular weight ($\overline{M}$v)

Determination of molecular weight of cellulose acetate did by viscometry. To measure the viscosity-average molecular weight ($\overline{M}$v), add five solutions with different concentration (1, 2, 3, 4 and 5 mg) of cellulose acetate and dissolved in 10 ml acetone. The efflux time for each solution was measured by Ostwald viscometer. $\overline{M}$v was determined with Mark-Houwink-Sakurada Eq. (3):

$$[\eta] = k\overline{M}_v^{\,a} \quad (3)$$

where $[\eta]$ = intrinsic viscosity; k and a are constants related to the polymer, solvent and temperature.

2.5 Characterization

FTIR spectra was performed on a Nicolet Avatar 360 using the KBr pellets method in the range 4000-400 cm^{-1} for pandan duri leaves, cellulose, and cellulose acetate powder from synthesis. The possible changes in the functional groups of cellulose acetate membranes at different composition of LiCl were investigated using Perkin Elmer, Frontier FT-IR Spectrometer Universal ATR sampling accessory in the range 4000-500 cm^{-1} with scanning resolution of 16 cm^{-1}. The conductivity of CA membranes were obtained using Elkahfi 100 in the range 0.05-9 V with maximum current 0.0035 A. The thickness of membranes were obtained using NSK micrometer. The conductivity of CA membrane was calculated by:

$$\sigma = \frac{\ln 2}{\pi t R} \quad (4)$$

where, σ = the membrane conductivity; t = the thickness of membrane; R = the membrane resistance. The crystallinity of CA membrane was obtained using a Rigaku Miniflex 600 Benchtop X-ray diffractometer equipped with CuK$_\alpha$ radiation at 40 kV and 15 mA in the range of 2θ = 2-90°. Crystallinity (X_c) of CA membrane was calculated by [11]:

$$X_c(\%) = \frac{A_c}{A_a + A_c}x100\% \quad (5)$$

where, Aa = area under amorphous peaks and Ac= area under crystalline peaks.

3 Results and discussion

In this work cellulose acetate was obtained by cellulose acetylation from pandan duri leaves using acetic anhydride as acetylating agent. Each step was characterized by FTIR to ensure the possible changes in functional groups of each product. Figure 1 shows the FTIR spectra obtained for (a) pandan duri leaves, (b) cellulose from pandan duri leaves and (c) cellulose acetate from pandan duri leaves. Figure 1a shows the band at 1734.15 cm^{-1} is attributed to C=O stretching of ester groups of hemicellulose and was no longer exist after the alkali treatment (Figure 1b) due to the cleavage of all ester groups of hemicellulose by alkali treatment [12, 13]. The absorbance peaks in the region 1511.91 cm^{-1} and 1253.28 reflect the stretching of the C=C aromatic and C-O-C (aryl-alkyl-ether) of lignin, respectively. These peaks slightly disappeared after pandan duri leaves dissolved in NaOCl, which indicated the absence of lignin [14].

Figure 1b shows the disappearance of the absorbance peaks in the region 1734.15 cm^{-1}, 1511.91 cm^{-1} and 1253.28 cm^{-1} suggest the successful of cellulose extraction from pandan duri leaves. The spectrum (c) in Figure 1 showing the presence of important cellulose acetate band at 1754.37 cm^{-1} (C=O ester), 1375.10 cm^{-1} (C-H bond in –OCOCH$_3$) and 1236.39 cm^{-1} (C-O-C of ester group of pure cellulose acetate) [15, 16]. In addition, the successful of acetylation process is shown by the absence of C=O bond (symmetric and asymmetric) in the region 1825-1760 cm^{-1} and 1700 cm^{-1} as indicated that the cellulose acetate is free of the unreacted acids (acetic anhydride and acetic acid) [17]. The calculated degree of substitution (DS) of cellulose acetate was obtained a value of 2.76.

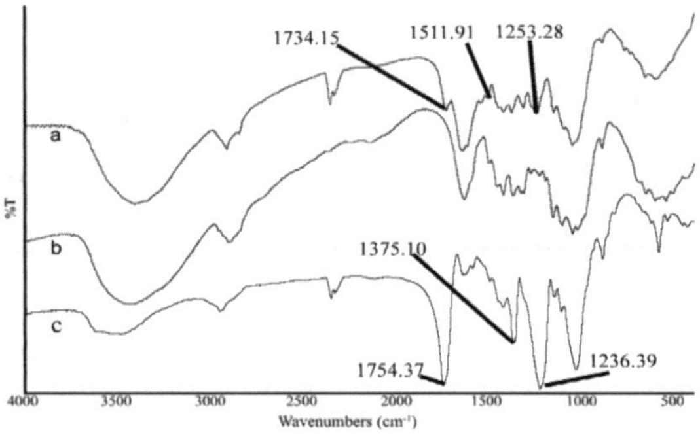

Figure 1. FTIR spectra of (a) pandan duri leaves, (b) cellulose of pandan duri leaves, and (c) cellulose acetate of pandan duri leaves.

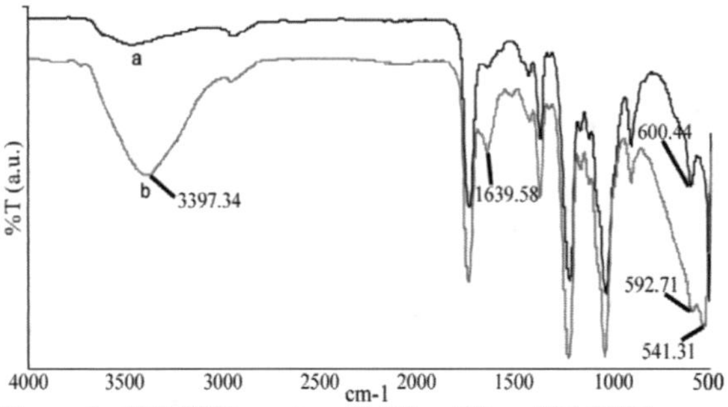

Figure 2. ATR-FTIR spectra of (a) Pure CA and (b) 65/35 CA/LiCl.

Table 1. Variation of conductivity of CA membranes under different salt concentrations.

CA/LiCl	Conductivity (S cm^{-1})
100/0	1.54 x 10^{-6}
80/20	2.52 x 10^{-3}
75/25	3.78 x 10^{-3}
70/30	1.10 x 10^{-2}
65/35	2.95 x 10^{-2}
60/40	1.65 x 10^{-2}

Table 1 shows the conductivity of cellulose acetate under different composition of LiCl with the thickness of membranes ranging from 0.001 to 0.035 cm. The calculated conductivity of pure cellulose acetate was obtained a value of 1.54 x 10^{-6} S cm^{-1} and the maximum conductivity value 2.95 x 10^{-2} S cm^{-1} was obtained at room temperature for 65/35 composition. The conductivity of cellulose acetate increase as LiCl concentration increase and decrease in 60/40 CA/LiCl composition. The conductivity of CA membrane increase due to the number of free ions (Li^+) and decrease due to lithium ion association in cellulose acetate membrane.

Figure 2 shows the ATR-FTIR spectrum of pure cellulose acetate and 65/35 CA/LiCl composition. It can be seen that the absorbance of hydroxyl functional group at 3397.34 cm^{-1} of 65/35 CA/LiCl composition is higher than pure cellulose acetate and peak at 1639.58 cm^{-1} (Figure 2b) as indicated HOH hydration due to the presence of lithium chloride in cellulose acetate membrane. In addition, the absorbance peaks at 541.31 and 592.71 cm^{-1} suggest the existence of lithium in 65/35 CA/LiCl composition that enhance the conductivity of its membrane [18]. The XRD pattern of 65/35 CA/LiCl shows in Figure 3, was obtained 2θ value around 17.25° and 23.02°. The low diffraction intensity indicate that the membrane was amorphous with the degree of crystallinity about 23.5%.

The addition of salt like lithium chloride in cellulose acetate membrane was decrease the crystallinity of cellulose acetate membrane due to the structural modification associated with cellulose acetate membrane. It was contribute to the increasing in the conductivity due to the increase of free volume of its membrane and the ions can freely move [19].

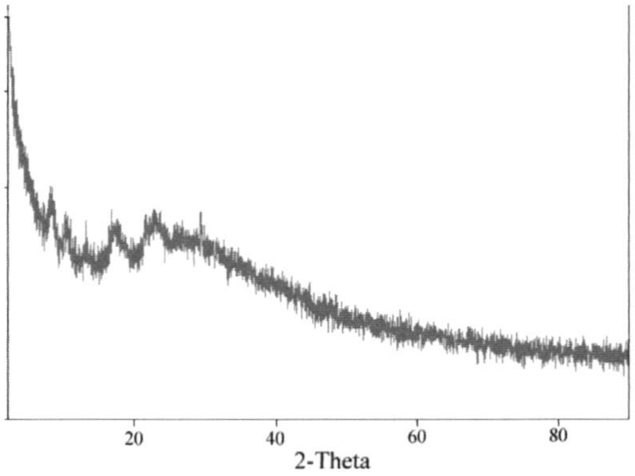

Figure 3. XRD pattern of 65/35 CA/LiCl.

For the used solvent system, acetone, the intrinsic viscosity was obtained a value of 119.72 mL/g and the Mark-Houwink-Sakurada constants are $K = 0.133$ and $a = 0.616$ [20]. The viscosity-average value molecular weight of cellulose acetate was 62,512 g mol^{-1}.

4 Conclusions

Cellulose acetate can be obtained by cellulose acetylation from pandan duri leaves. The highest conductivity of cellulose acetate membrane was obtained for 65/35 CA/LiCl composition with a value of 2.95 x 10^{-2} S cm^{-1}. For the future work, the optimization of acetylation and addition of polymer like PEG or PVA are needed to increase the conductivity and mechanical properties of membrane.

Acknowledgments

This work was supported by DIPA UNY grant.

References

1. M. Rikukawa, K. Sanui, Prog. Poym. Sci., **25**, 1463-1502 (2000)
2. M. Lowe, S. Takuoka, T. Trigg, G. Gareffi, *Lithium-ion Batteries for Electric Vehicles: The U.S. Value* Chain (CGGC, NC, 2010)

3. J. S. Liu, W. S. Li, X. X. Zuo, S. Q. Liu, Z. Li, Journal of Power Source, **226**, 101-106 (2013)
4. D. Alves Cerqueira, G. Rodrigues Filho, C. da Silva Meireles, Carbohydrate Polymers, **69**, 578-582 (2007)
5. J. Cai, P. Fei, Z. Y. Xiong, Y. J. Shi, K. Yan, H. G. Xiong, Carbohydrate Polymers, **92**, 11-18 (2013)
6. Parameswara Rao Chinnam, Hanjun Zhang, Stephanie L. Wunder, Electrochimica Acta, **170**, 191-201 (2015)
7. M. Selvakumar, D. Krishna Bhat, Journal of Applied Polymer Science, **110**, 594-602 (2008)
8. N. A. Razuki, A. W. Aziz, N. S. A. Satar, Noor Haida Mohd Kaus, Int. J. Electroactive Mater, **3**, 38-45 (2015)
9. W. Giesen, S. Wulffraat, M. Zieren, L. Scholten, *Mangrove Guide Book for Southeast Asia* (FAO and Wetlands International, The Netherlands, 2006)
10. R. M. Sheltami, I. Abdullah, I. Ahmad, A. Dufresne, H. Kargarzadeh, Carbohydrate Polymers, **88**, 772-779 (2012)
11. S. Gaidukov, I. Danielenko, G. Gaidukova, International Journal of Polymer Science, **2015**, 1-8 (2015)
12. A. Alemdar, M. Sain, Bioresource Technology, **99**, 1664-1671 (2008)
13. X. F. Sun, F. Xu, R. C. Sun, P. Fowler, M. S. Baird, Carbohydrate Research, **340**, 97-106 (2005)
14. B. Xiao, X. F. Sun, Run Chang Sun, Polymer Degradation and Stability, **74**, 307-319 (2001)
15. A. M. Das, A. A. Ali, M. P. Hazarika, Carbohydrate Polymers, **112**, 342-349 (2014)
16. C. N. Saikia, F. Ali, T. Goswami, Anil C. Ghosh, Industrial Crops and Products, **4**, 233-239 (1995)
17. X. F. Sun, R. C. Sun, J. X Sun, Bioresource Technology, **95**, 343-350 (2004)
18. Jean-Joseph Max, Camille Chapados, J. Chem. Phys., **115**, 2664 (2001)
19. A. Ahmad, M. Y. A. Rahman, S. P. Low, H. Hmazah, ISRN Materials Science, **2011**, 1-7 (2011)
20. K. Kamide, T. Terakawa, Y. Miyazaki, Polymer Journal, **11**, 285-289 (1979)

MATEC Web of Conferences **64**, 05001 (2016) DOI: 10.1051/matecconf/20166405001
ICCMP 2016

Transparent Aerogels with High Mechanical Strength Composed of Cellulose-silica Cross-linked networks

Qing Zhou, Yong Shen , Sufen Ai, Bin Liu and Yingmin Zhao

Aerospace Institute of Advanced Materials & Processing Technology, CHN

Abstract. We prepared monolithic transparent composite aerogels with high mechanical strength composed of cellulose-silica interpenetrating polymer network (IPN) structures. Cellulose-silica areogels were obtained by dissolving cellulose in silica hydrogels, followed by gelling, aging and drying using supercritical drying. The composite nanonetworks of cellulose-silica areogels were demonstrated with scanning electron microscope (SEM). The density of cellulose-silica aerogels is between 20 and 38 kg/m^3. The surface area of aerogels was between 300 m^2/g to 500 m^2/g.

1 Introduction

Aerogels are highly porous nanostructured materials usually prepared by sol-gel processing followed by supercritical drying whose final stage involved extracting the pore-filling solvent with liquid CO_2 [1-6]. Aerogels which are characterized by large internal surface area and large open pores, are promising candidates for various advanced applications such as thermal insulation, catalysts/catalyst supports, acoustics, and gas filters [7-15].

Currently, the ever worsening problems of fossil energy depletion and global warming have become a major challenge and survival threat to human being, and have thus greatly boosted the research efforts in development of clean alternative energy and emission control of global warming gases. However these effortswould prove insufficient if not accompanied with energy conservation practices. One of the major energy saving practices is thermal insulation, e.g., reducing fuel and electricity usages through better insulating boilers and buildings, respectively. In this situation, aerogels are prospective to be used in the glazing of green buildings because of their near transparency and low thermal conductivity.

Though monolithic transparent aerogels can be facilely prepared from silica aerogels, they suffer from the poor mechanical properties due to the weak chemical bonding among silica nanoparticles. One solution is to fabricate cellulose aerogels [16-19] with interconnected open-porous network structures. However, they ' re light-tight, and the preparation process usually involves the utilization of toxic isocyanates. Furthermore, neat aerogels including silica aerogels and cellulose aerogels shrink during production and the shrinkage cracking makes it difficult to form monolithic aerogels. Therefore, a couple of scientific studies have been initiated aiming at the preparation of silica aerogels [20-23].

In this paper, we reported a novel method to fabricate composite cellulose-aerogels with a interpenetrating polymer network (IPN) structure. We used the sol–gel synthesis method toward nanostructured silica, which typically started from tetramethoxysilane (TMOS). The resulting composite aerogels were dried with supercritical CO_2 to give cellulose–silica aerogels which were lightweight, transparent and with good mechanical properties. Ionic liquid was used as a solvent to dissolve native cellulose. The native cellulose was esterifed to make it dissolved well in silica sol solution. The cellulose-silica composite areogels can be used as glazing in energy-saving building.

2 Experimental section

2.1 Preparation

2.1.1 Materials

Tetramethoxysilane (TMOS) were purchased from Sigma-Aldrich. Fibrous native cellulose (Ashless pulp. Advantec, Japan), n-Butyric anhydride, hydrochloric acid (37%), formic acid and ethanol (EtOH) were obtained from Fluka. Ionic liquid (AmimCl) was purchased from Sigma-Aldrich. All reagents were used as received unless otherwise noted.

2.1.2 Cellulose modified

Native cellulose was firstly dissolved in ionic liquid at room temperature with stirring to form a cellulose solution, and subsequently esterifed by Butyric anhydride with catalysts (ammonia), and three kinds of esterifiable celluloses named P1, P2, P3 which differed in degree of esterification could be obtained.

2.1.3 Wet gel preparation

Cellulose-silica hydrogels were fabricated according to thefollowing procedure:TEOS and EtOH were mixed in a beaker. Then HCl (0.038M) was added and the mixtures were stirred for 1.5h. A solution of esterified cellulose in ethanol was then added to the mixture and stirred for 5min. The wet cellulose-silica hydrogels were obtained after standing for 24h.

2.1.4 Drying method

The cellulose-silica hydrogels were solvent-exchanged with EtOH for 3 times, and then subjected to supercritical carbon dioxide drying. The autoclave is closed and filled with liquid carbon dioxide (CO_2) until the samples are completely immersed in the liquid. The EtOH iscompletely removed after several times washing with pure CO_2.

2.2 Characterization

FTIR measurements were performed on a Perkin-Elmer series 2000 instrument. Sample wafers consisted of 105 mg dry KBr and ca. 1 mg aerogel sample.

^{1}H MAS NMR spectra was obtained on a Bruker DSX-400 spectrometer at 400 MHz.

Densities of the aerogels were calculated by weighting the samples and measuring the volumes.

Scanning electron microscopy (SEM) was conducted on a Hitachi S-4800 microscopy.

The specific surface area of aerogels was measured with the BET method from the nitrogen adsorption in a Coulter SA3100 analyser. The analysis was done for relative vapor pressure of 0.05 to 0.3.

Lard and strain tests of three-point flexural bending method of the composite aerogels were performed on an EZ-test (DDL-100, China) with a deformation rate of 1 mm/min. The aerogels of a size of 80 mm length and a square cross section with an edge length of 20 mm.

3 Results and discussion

Fig. 1 shows the esterification of native cellulose. To distinguish the different degree of esterification, we called the modified cellulose P0, P1, P2 and P3, with P0 representing the original cellulose.

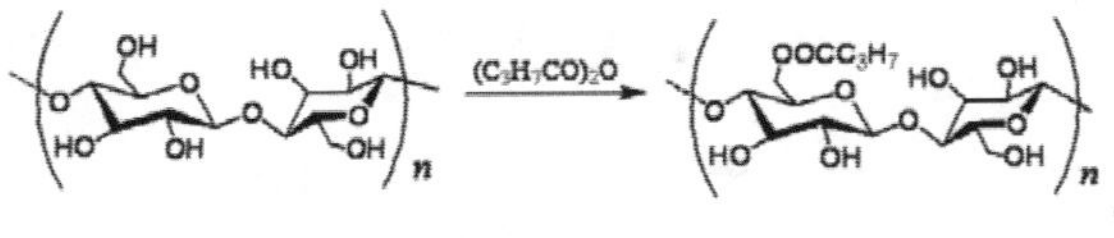

Figure 1. Esterification of native cellulose.

The esterification pruducts of cellulose were characterized by FT-IR and ^{1}H-NMR spectra. Fig. 2 gives the FT-IR spectra of the original cellulose P0 and the esterifiable products P1, P2 and P3, respectively. Compared to the FT-IR spectra of P0, the appearance of absorption located at 1745 cm^{-1} and 1167 cm^{-1} suggests the successful esterification of cellulose. The absorption located at 2968 cm^{-1} and 2870 cm^{-1} can be ascribed to the stretch vibration of methyl group while the absorption located at 2940 cm^{-1} and 1420 cm^{-1} are attributed to methylene group. These results further indicate the successful introduction of alkyl group due to esterification. Moreover, the higher absorbance at 1745 cm^{-1} of P3 compared to that of P2 demonstrates the higher reaction conversion of P3. P1 has the lowest reaction conversion.

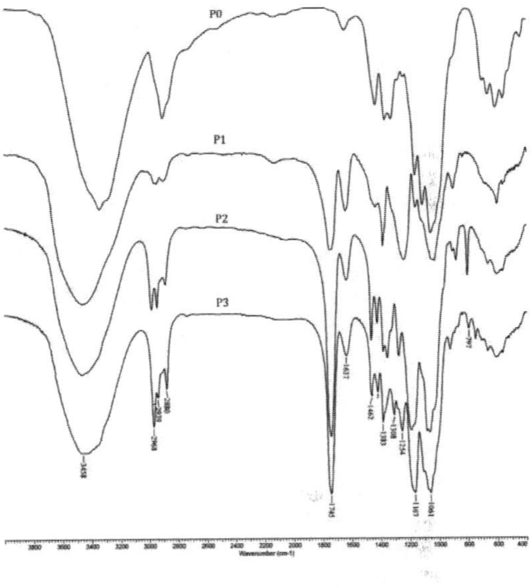

Figure 2. FTIR of native cellulose and esterified cellulose.

The structures of P1, P2 and P3 were further characterized with ^{1}H-NMR spectra. As shown in Fig. 3, the broad peak at 2.2 ppm can be assigned to the resonance of methylene group adjacent to carboxyl group and the broad peak at 1.5 ppm can be attributed to another methylene group. The resonance of methylene group together with that of methyl group at 0.87 ppm demonstrated the successful esterification of cellulose P0. It is worth pointing out that it is difficult to quantitatively calculate the reaction conversion of P1, P2 and P3 due to their poor solubility in DMSO. Sample P3 was chosen to continue the following experiment because of its good solubility in silica solution.

^{13}C-NMR spectrum is another powerful tool in characterization of organic compounds. Fig. 4 gives the ^{13}C-NMR spectrum of P3. The three peaks located at 172.4 ppm, 171.6 ppm, 171.2 ppm can be assigned to the carbon of the ester group. The peak at 35.0, 17.7 ppm and 13.3 ppm can be assigned to the carbon of methylene and methyl group, respectively. The peaks located between 56 ppm and 104 ppm are attributed to the carbon of glucose unit. P1 and P2 give consistent and similar ^{13}C-NMR spectra with P3. The ^{13}C-NMR spectra further demonstrate the successful esterification of native cellulose.

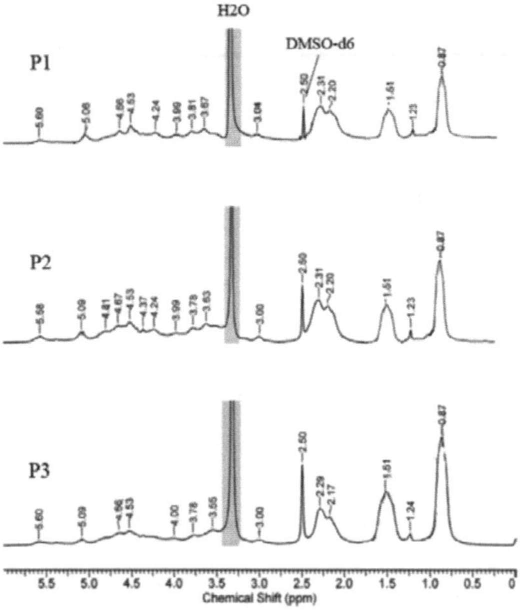

Figure 3. ^{1}H-NMR of P1, P2 and P3.

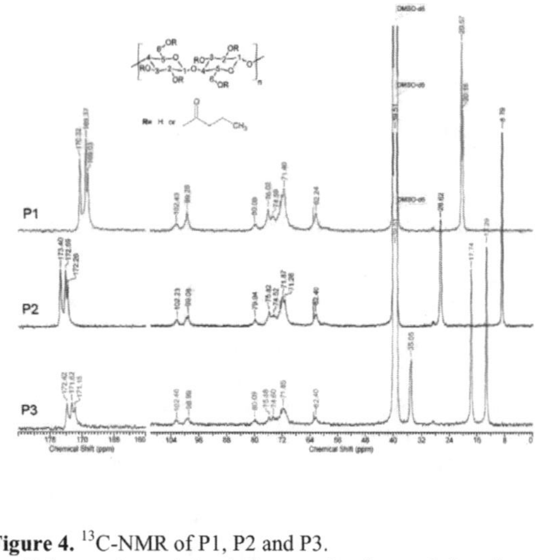

Figure 4. ^{13}C-NMR of P1, P2 and P3.

Scheme 1 explained the mechanism of formic acid catalyzing hydrolysis-condensation of TMOS and silica hydrogels. The whole process took place in ionic liquid. Fig. 5 illustrates the preparation of the cellulose-silica aerogels. The silica hydrogels were firstly formated, then cellulose molecular aggregated around the silica network to form a interpenetrating polymer network (IPN) structure hydrogel, at last the composite hydrogel was supercriticallly dried by liquid CO_2 to obtain silica-cellulose composite aerogels. Fig. 6 shows the pictures of the cellulose-silica hydrogels and aerogels with cellulose concentrations vary from 0.75 wt.% to 10 wt.% respectively. All the hydrogels are transparent without obvious difference, as can be seen from the pictures in Fig. 6. However, the transparency of aerogels decreases as the concentration of cellulose increases. Table 1 summarizes the sample compositions and densities. The sample densities increases from 20 to 38 kg/m^3 as the content of cellulose changes from 0.75 wt.% to 10 wt.% in concentration.

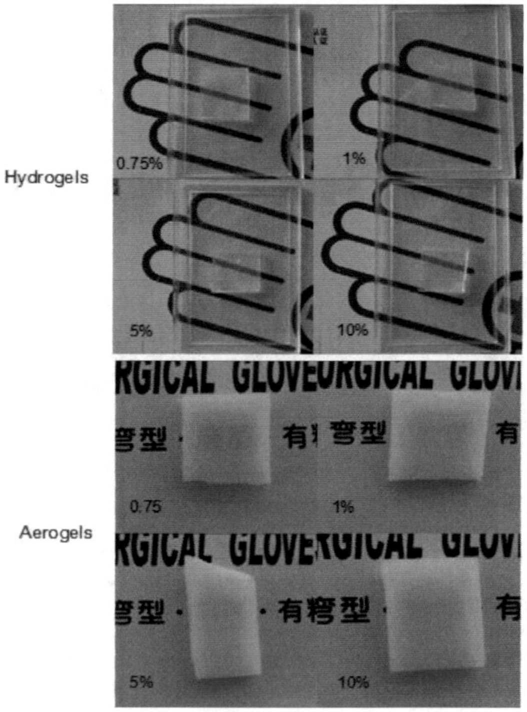

Figure 6. Pictures of cellulose-silica hydrogels and aerogels with different cellulose contents.

Table 1. Densities of aerogels with different cellulose concentrations.

	Cellulose/wt.%	TEOS/wt.%	Density/kg m^{-3}
IPN1	0.75	8	20
IPN2	1	8	24
IPN3	5	8	29
IPN4	10	8	38

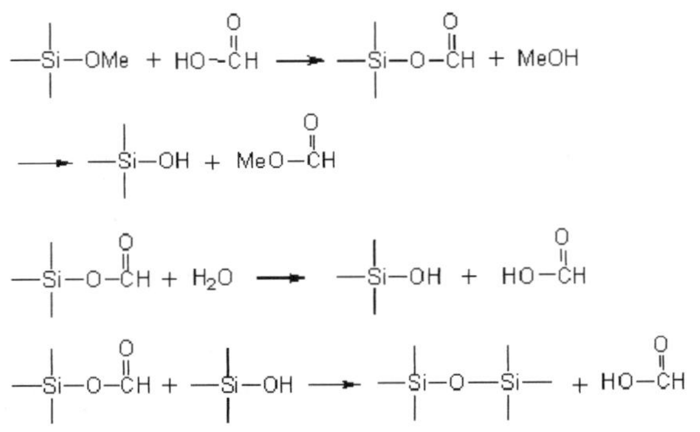

Scheme 1. The mechanism of TMOS hydrolysis-condensation catalyzed by formic acid.

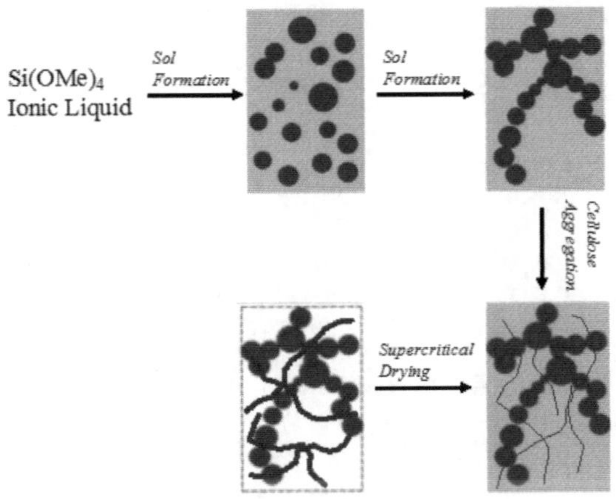

Figure 5. The schematic preparation of Cellulose-silica aerogels.

The microstructures of cellulose-silica aerogels are observed by SEM. These observations indicate that instead of being homogeneous, cellulose particles deposit around silica to form a stronger network. The SEM pictures furthermore demonstrates that the size of composite nanoparticles varies from 50 nm to 1 μm. When the concentration of the cellulose solution is 0.75 wt.%, the diameter of aerogel paricles is about 50 nm. When the concentration of the cellulose solution is 1 wt.%, the diameter of aerogel paricles is about 300 nm; When the concentration of the cellulose solution is 5 wt.%, the diameter of aerogel paricles is about 700 nm; When the concentration of the cellulose solution is 10 wt.%, the diameter of aerogel paricles is about 1 μm. cellulose Those particles are connected by "necks" formed by dissolution and reprecipitation of cellulose during aging. However, the micrographs doesnot show the mesopores which were characterized by the BET method from the nitrogen adsorption.

Figure 7. (a) SEM picture of a composite aerogel with 0.75 wt.% cellulose. (b) SEM picture of a composite aerogel with 1 wt.% cellulose. (c) SEM picture a composite aerogel with 5 wt.% cellulose. (d) SEM picture a composite aerogel with 10 wt.% cellulose.

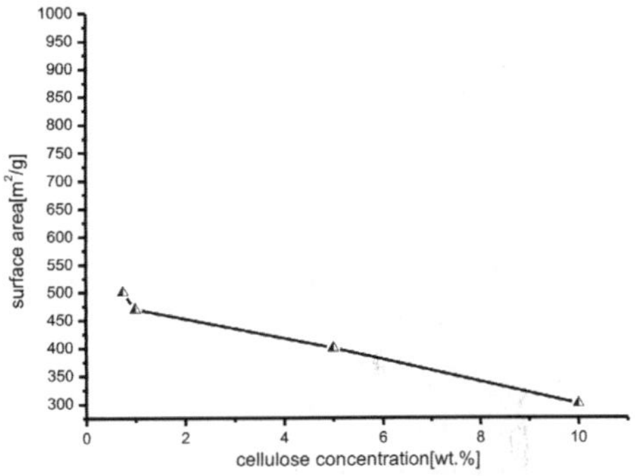

Figure 8. BET-schema for cellulose-silica composite aerogels.

Fig. 8 exhibits the surface area of silica-cellulose composite aerogels produced by different concentration of cellulose solution. The surface area of composite aerogels with 0.75 wt.% cellulose is about 525 m^2/g; The surface area of composite aerogels with 1.00 wt.% cellulose is about 480 m^2/g; The surface area of composite aerogels with 5.00 wt.% cellulose is about 430 m^2/g; the surface area of composite aerogels with 10.00 wt.% cellulose is about 300 m^2/g. The figure of the BET indicates that the surface area of the composite aerogels is inversely proportional to the cellulose concentration, which is consistent with the results from SEM images.

Fig. 9 shows the load-strain curves of four representative composite monoliths on the way to their respective rupture points. The most dramatic improvement of the composite aerogels is in the strength of the new material. It's a test called three point flexural bending method. It takes more than 10 times higher load to break a aerogel monolith with density 0.038 g/cm^3 (12 kg) than to break a aerogel monolith with density 0.020 g/cm^3 (12 kg). The least dense sample is linearly elastic, while the more dense samples behave as nonlinear elastic.

When the density of composite aerogels reaches to 0.038 g/cm^3, the rupture point is about 12 kg that it is strong enough to be carried on.

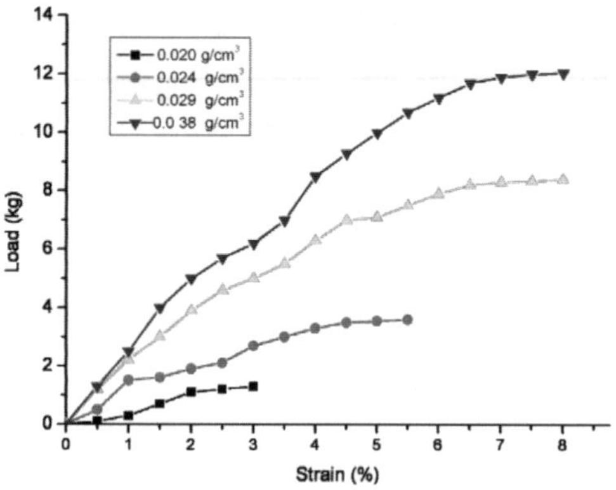

Figure 9. Load-strain curves for the four composites of table 1.

4 Conclusions

High mmechanical strength and flexible cellulose-silica aerogels composed of Cellulose-silica Cross-linked networks were fabricated by sol-gel chemistry using supercritical drying. It's possible to build complex 3D-structures at nanometer range between silica and cellulose. Ionic liquid was used as a solvent to dissolve native cellulose. The native cellulose was esterifed to make it dissolved well in silica sol solution. The molecules of cellulose deposited around the network of silica and reinforced the structure. The cross-linked porous structures can be shown in SEM. Their densities were between 20 and 38 kg/m^3 with concentrations of cellulose between 0.75 wt.% and 10 wt.%. The surface area of aerogels was between 300 m^2/g to 500 m^2/g. When the density of composite aerogels reaches to 0.038 $g\ cm^{-3}$, the rupture point is about 12 kg. In this way, it's possible to fabricate glazing used in green building in the future.

Acknowledgments

This work is supported by Chinese state commission of science and technology for national defense industry and it's managed by the aerospace Institute of advanced materials & processing technology, has sponsored this work.

References

1. N. H. Sing, U. Schubert, Angew. Chem. 1998, 110, 22 – 47; Angew. Chem. Int. Ed., **37**, 22 – 45, (1998).
2. A. C. Pierre, G. M. Pajonk, Chem. Rev., **102**, 4243 – 4265, (2002).
3. A. E. Aliev, J. Y. Oh, M. E. Kozlov, A. A. Kuznetsov, S. L. Fang, A. F. Fonseca, R. Ovalle, M. D. Lima, M. H. Haque, Y. N. Gartstein, M. Zhang, A. A. Zakhidov, R. H. Baughman, Science, **323**, 1575 – 1578, (2009).
4. J. L. Mohanan, I. U. Arachchige, S. L. Brock, Science, **307**, 397 – 400, (2005).
5. M. A. Worsley, P. J. Pauzauskie, T. Y. Olson, J. Biener, J. H. Satcher, Jr., T. F. Baumann, J. Am. Chem. Soc., **132**, 14067 – 14069, (2010).
6. I. U. Arachchige, S. L. Brock, J. Am. Chem. Soc., **128**, 7964 – 7971, (2006).
7. J. Fricke, Aerogels, Springer-Verlag, Berlin (1986).
8. R. J. Ayen and P. A. Iacobucci, Reviews in Chemical Engineering **5**, 157(1988).
9. M. Schneider and A. Baiker, Catal. Rev.-Sci. Eng. **37**, 515 (1995).
10. Pajonk, G. M., Catal. Today, **37**, 3(1999).
11. Morris, C. A.; Anderson, M. L.; Stroud, R. M.; Merzbacher, C. I.; Rolison, D. R., Science, **284**, 622(1999).
12. Novak, B. M.; Auerbach, D.; Verrier, C., Chem. Mater., **6**, 282(1994).
13. Jones, S. M. J., Sol-Gel Sci. Tech., **40**, 351(2006).
14. Junga, S. B.; Parka, S. W.; Yanga, J. K.; Park, H. H.; Kimb, H., Thin Solid Films, **580**, 447–448(2004).
15. Tsutsui, T.; Yahiro, M.; Yokogawa, H.; Kawano, K.; Yokoyama, M.; AdV. Mater., **13**, 1149(2001).
16. Fischer F, Rigacci A, Pirard R, Berthon-Fabry S, Achard P, Polymer, **47**, 7636, (2006)
17. Innerlohinger J, Weber HK, Kraft G, *Macromol Symposia,* **244**, 126, (2006).
18. Liebner F, Haimer E, Potthast A, Loidl D, Tschegg S, Neouze M-A, Wendland M, Rosenau T, Cellulosic aerogels as ultra-lightweight materials. Part 2: synthesis and properties, **63**, 3, (2009).
19. Liebner F, Haimer E, Wendland M, Neouze M-A, Schlufter P, Miethe P, Heinze T, Potthast A, Rosenau T, Macromol Biosci **10**, 349, (2010).
20. Aljaberi A, Chatterji A, Shah NH, Sandhu HK, Drug Development and Industrial Pharmacy, **35**, 1066, (2009).
21. Gill RS, Marquez M, Larsen G, Micropor Mesopor Mater, **85**, 129(2005).
22. Goncalves G, Marques PAAP, Trindade T, Pascoal Neto C, Gandini A, J Colloid Interface Sci, **324**, 42, (2008).
23. Tanaka K, Kozuka H, J Sol-Gel Sci Technol, **32**, 73, (2004).

Estimation of NO_x Emission from Fossil Fuel Consumption in China for the Period 1980-2012

Yu Du[1,2] and Xixiang Sun[1]

[1]*School of Management, Wuhan University of technology, Wuhan 430070, China*
[2]*Agriculture bank of China Shandong Branch, Jinan 250000, China*

Abstract. As the largest consumer of fossil fuel, China NOx emission from energy consumption has become a hotspot for studies. In this study, emission inventory was used to analysis the historical variation of NOx emission in china for the period 1980 to 2012. Results indicate that NOx mission of China has increased from 446.45×104t to 2499.72×104t during 1980-2012, electricity, industry and transportation was the main contributor and NOx emission has increased from 145.04×104t, 206.75×104t, 39.44×104t to 1311.6×104t, 479.46×104t, 576.31×104t respectively; from the spatial pattern, high total NOx emission with mainly concentrated in the north of China and the high emission intensity mainly concentrate in eastern China.

1 Introduction

It is now well known that NO_x emissions play an important role in acidification, haze, summer photochemical smog and the increase of tropospheric ozone level [1]. Fossil fuel combustion is identified as one of the main contributors to the escalation of air pollution such as NO_x [2]. China is the largest energy consumer in the world and have contributed amount of various emission during past three decades because of its rapid industrializations and urbanizations. Although China has implement strict environment policy in recent years, NO_x emission still keeps an increase trend all the way. Data from satellite measurements indicated that NO_x vertical column density have increased significantly in East Asia [3].

A systematic emission inventory is helpful for environmental policy maker to understand the essential information of emission source and control air pollution effectively. Several national and provincial level NO_x emission inventories have been presented by several researchers. For instance, D.G.Streets(2000) used emission factor estimate China provincial NO_x emissions of 1995 and forecasted emissions for the year 2020[4]; Tian Hezhong(2001) build up 1980-1998 NO_x emissions inventory of China on the basis of historical energy consumption data and NO_x emission factors[5]; Qing Lu(2013) took use of emission inventory estimate the NO_x emission of Pearl River Delta region and found NO_x emissions went up consistently during 2000-2009 except for a break point in 2008[6]

Yun Shi *et al.* (2014) established a NOx emissions inventory of China from for the period 2000-2010[7]. These study focus more on short period or annual emission and pay no more attention to spatial distribution variation of NO_x emission. As a country with the fast developing speed in the world, China NO_x emission changed greatly over the past years, in this paper we established a bottom-up NO_x emission inventory of China from 1980 to 2012 for different economic sectors and different fuel styles, then, based on the provincial fossil fuel consumption of 2012, we analyses spatial distribution characteristic of NO_x emission in provincial level.

2 Data and methods

Following the approach of Tian Hezhong(2001)[5] and Qing Lu(2013)[6] NOx emission source were classified into seven categories: power plants and heating, industry, construction, transportation, commercial, residential consumption and other resources. Ten fuel types such as coal, coke, crude oil etc, are considered and their emission factors are different for different period. NO_x emission was estimated by an emission factor approach which was implemented by the Equation below:

$$Q_{i,j,f}^{N}(t)=K_{i,j,f}^{N}(t)\times F_{i,j,f}(t)\times(1-P_{i,j,f}^{N}) \quad (1)$$

$$Q_{i,j}^{N}=\sum_{f}\sum_{j}Q_{i,j,f}^{N}(t) \quad (2)$$

where Q^N is the emission of NO_x, K^N is the emission factors of NO_x weighted by N_2O, F is the fuel consumption, P is the average removal efficiency of NO_x, i, j, f, t represent different province (municipality or Autonomous Region, except Hong Kang, Macau and

Taiwan due to the lack of available data), economic sectors, fuel styles and years respectively.

Emission factor is the basis for inventory, in order to improve the precise and systematic of estimation, we performed an extensive literature reviews on NO_x emission factors and adopted the emission of Tian Hezhong[5] and Zhang Chuying[8]. Considering the technology development of electricity and industry in fact, new factors was used in the two sectors from the year of 2005 and new emission factors of 2005 was marked with * . The unit for all emission factors was expressed by kg/t except for natural gas and coal gas expressed by $kg/10^4{\cdot}m^3$. All emission factors used in this paper were listed in tab.1. Fuel consumption data of different sectors came from national and provincial energy balance table (physical quantity) of National Statistic Yearbook (1985-2013).

Table 1. NOx emission factors for seven sectors and different fossil fuels of China (kg/t).

	Coal	Coke	Crude oil	Gasoline	Kerosene	Diesel	Fuel oil	LPG	Refinery dry gas	Natural Gas[1]	Coal gas[1]
electricity	9.95 6.58*		7.24 5.84*	16.7 5.84*	21.2	7.40	10.06 5.84*	3.74 2.1*	0.75	40.96	13.53
industry	7.5 4*	9.0 4.8*	5.09	16.7	7.46	9.62	5.84	2.63	0.53	20.85	9.5
construction	7.5	9.0		16.7	7.46	9.62	5.84	2.63	0.53	20.85	9.5
transportation	7.5	9.0	5.09	21.2	27.4	36.25	18.1	18.1		20.85	
Commercial	3.75	4.5	3.05	16.7	4.48	3.50	1.58	1.58	0.32	14.62	7.36
Residential	1.88	2.25	1.70	16.7	2.49	1.95		0.88	0.18	14.62	7.36
others	3.75	4.5	3.05	16.7	4.48	3.50	1.58	1.58	0.32	14.62	7.36

3 Results

3.1 Temporal and sector variation of China NO_x emission

Fig. 1 and Fig. 2 illustrate the emissions and contributions of various sources. It can be found that the total NO_x emission from various sources has increased from $446.45{\times}10^4$t to $2499.72{\times}10^4$t and increased 4.6 times from 1980 to 2012, kept the rapid growth in recent 10 years because of energy consumption increase. Electricity was the absolutely dominated contributor of NO_x all the way in the past three decades, NO_x emission from electricity has increased from $145.04{\times}10^4$t to $1311.6{\times}10^4$t and the emission ratio of electricity has increased from 32.48% to 52.47% (Fig. 2). Industry was the second contributor after electricity, NO_x emission from industry has increased from $206.75{\times}10^4$t to $479.46{\times}10^4$t and increased about 1.31 times than that of 1980, however the emission ratio of industry has decreased from 46.31% to 19.18%. Transportation was the fast growing source, NO_x emission has increased from $39.44{\times}10^4$t to $576.31{\times}10^4$t and increased about 13.61 times than that of 1980. NO_x emission from other sectors, including construction, residential, others, was little and takes up less than 5% of emission only.

Although NO_x emission proportion of coal consumption have declined in recent years, coal still was the main contributor of NO_x and NO_x emission of coal combustion has increased from $328.63{\times}10^4$t to $1610.45{\times}10^4$t from 1980 to 2012(Fig.3). NO_x emission of diesel and gasoline combustion has increased from $20.94{\times}10^4$t, $18.50{\times}10^4$t to $430.70{\times}10^4$t $152.53{\times}10^4$t because of rapid increased motor vehicle. Coke was the important component of steelmaking; NO_x emission from coke has increased from $38.54{\times}10^4$t to $181.11{\times}10^4$t and had a stable status in NO_x emission. Other fuel style contributed little NO_x and took no more than 3% of NO_x emission.

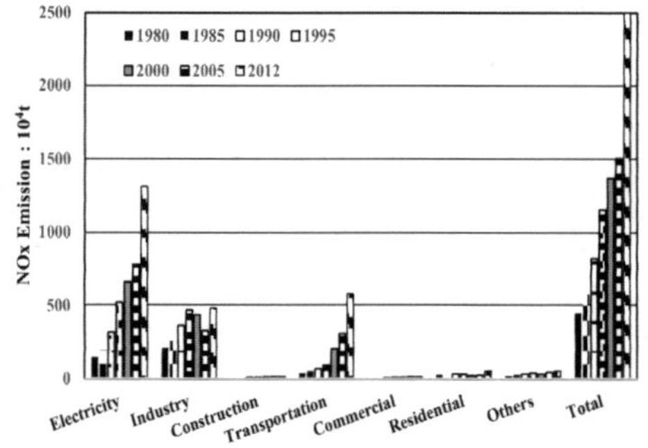

Figure 1. NO_x emission of main sectors from 1980-2012.

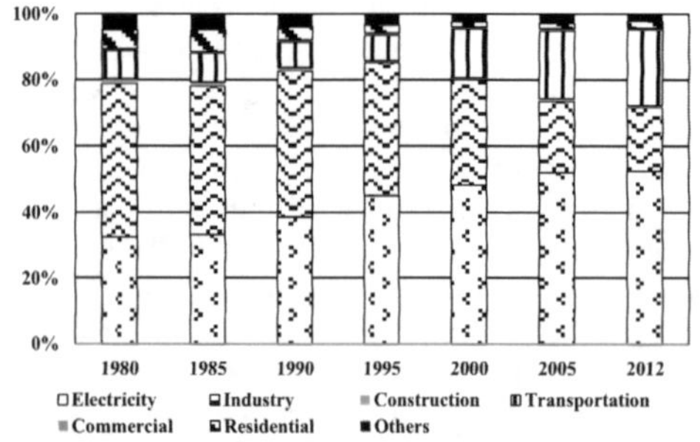

Figure 2. The contribution of main sectors to NO_x emission from 1980-2012.

There wasn't official NO_x emission statistic data in China until 2006, in order to verify the data validity of this paper, we contrasted the NO_x emission of relevant literature and listed them in fig.4, among them, data of 2012 was from National Environmental Statistic year book (2013), data of 1980, 1985, 1995 was from Tian

DOI: 10.1051/matecconf/20166406001

hezhong(2005) [5], data of 1990 from Wang wenxing(1996) [9], data of 2000 and 2005 was from Shi Yun (2014) [7]. Although data source and method was different, NO_x emission had same increase trend and had a little difference both in this paper and other researchers, which was less than 10% except data of 2000 and 2005. So we thought the estimation result of this paper was reliable.

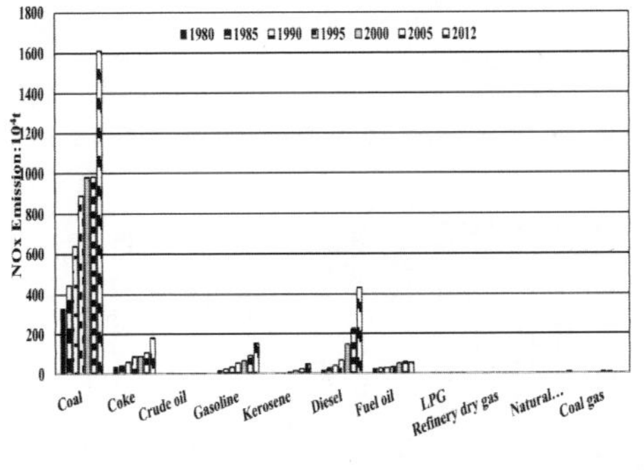

Figure 3. NO_x emission from various fossil fuels (1980-2012).

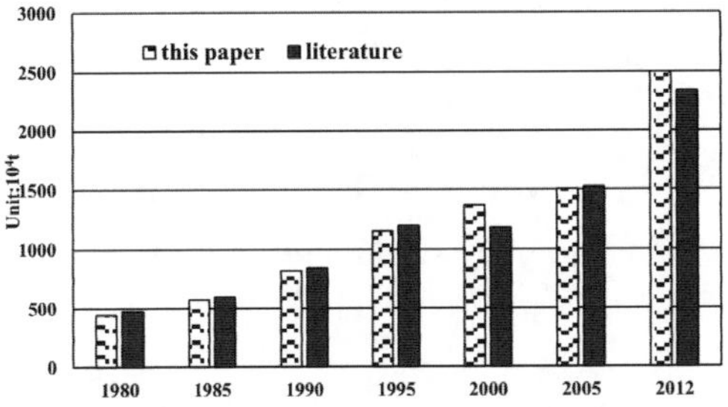

Figure 4. Data contrast from various data source (1980-2012).

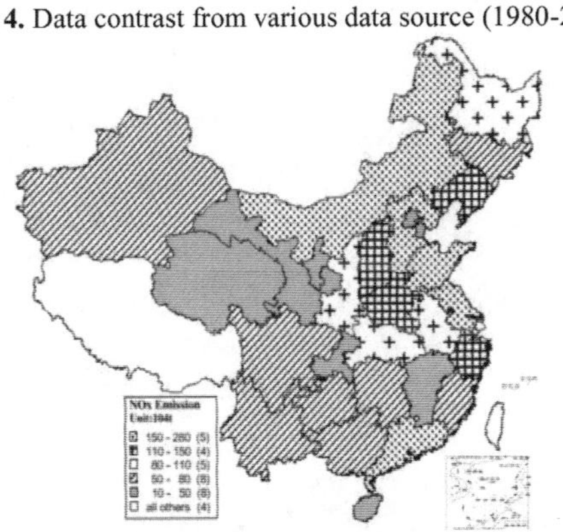

Figure 5. Spatial distribution of NOx emission from fossil fuel combustion for the year 2012.

3.2 Spatial distribution of China NO_x emission

There are great disparities in total NO_x emission from different provinces and mainly lie in north of China for the year 2012 (Fig. 5). Shandong, Inner Mongolia, Jiangsu, Guangdong and Hebei province had a high NO_x emission and their NO_x emission was 279.66×10^4t, 208.74×10^4t, 201.27×10^4t, 177.68×10^4t, and 171.64×10^4t respectively. Provinces with low NO_x emission was relative disperse and showed a polarization phenomenon in emission scale, these province including Qinghai, Hainan, Beijing, Chongqing, Gansu, Ningxia, Jiangxi, Tianjin and their NO_x emission was 10.5×10^4t, 13.61×10^4t, 36.59×10^4t, 39.52×10^4t, 44.15×10^4t, 44.17×10^4t, 45.55×10^4t, 46.02×10^4t respectively in 2012.

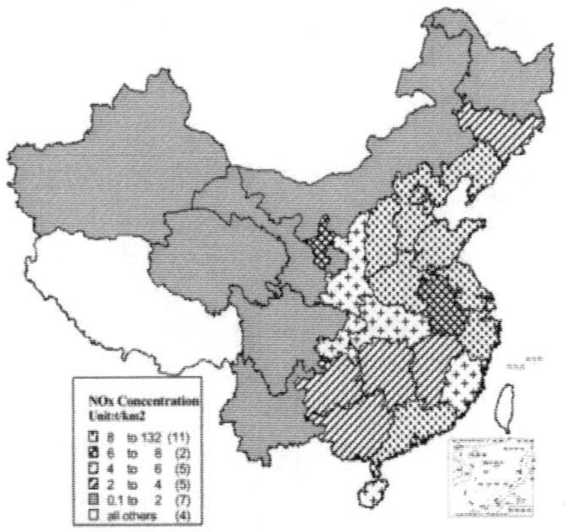

Figure 6. Spatial distribution of NOx intensity from fossil fuel combustion for the year 2012.

Due to difference in land area, the spatial distribution of NO_x intensity took a distinct feature of zonality in eastern, middle and western China (Fig.6). Shanghai, Tianjin and Beijing, Jiangsu, Shandong and Zhejiang, had the highest emission density of NOx due to their concentration of population and industry polluted area, their NOx intensity reached to 131.8t/km^2, 40.72 t/km^2, 21.28 t/km^2, 19.62 t/km^2, 18.01 t/km^2 and 12.05 t/km^2 respectively, which were similar to data via satellite monitoring[1]. NOx intensity of western China such as Qinghai, Xinjiang, Gansu, Sichuan, Yunnan, Inner Mongolia, Heilongjiang provinces was low owing to their broader land area, whose NOx intensity only 0.15 t/km^2, 0.43 t/km^2, 0.97 t/km^2, 1.65 t/km^2, 1.66 t/km^2, 1.74 t/km^2, 1.78 t/km^2 respectively.

3.3 NO_x speciation by provinces

Fig.7 showed the NO_x speciation by provinces for the year of 2012. The profile for provinces were somewhat similar, mainly electricity and industry. However, some differences can be seen, for example, transportation was predominant in Shanghai, Beijing and Hainan, industry was predominant in Sichuan, Chongqing, Hubei and Hunan province. The contribution of economic sector indicated that NOx emission control policy should be suited to local condition.

4 Discussion and conclusions

Using the inventory method on based emission factor, we estimated the NO_x emission of China from 1980 to 2012 and provincial distribution for the year 2012 for the first time. Results indicated that NO_x had increased 4.6 times during the past 32 years and had an apparent upward trend in recent years, because of rapid fossil fuel consumption. Electricity has exceeded industry and become the main contributor of NO_x emission. Because of its rising ratio in NO_x emission, electricity is the key sector for NO_x emission control in the future. Industry was the important contributor for its gigantic total NO_x emission although its emission ratio has declined, improving industrial energy consumption efficiency and enforcing pollution control was the only way for NO_x emission. Transportation was the fast growing contributor for NO_x emission and the public transport system should be as a priority measures to reducing the growth of NOx emission. From the perspective of energy structure, coal as the main energy source for NO_x emission and the promotion of green energy would be significant to decrease the NO_x emission. The spatial patterns of NOx emission in China energy consumption were primary concentrated in eastern of China because of their dense population and industrialization.

A long time series NO_x emission inventory was developed for the first time. Although different in data source and the method, the estimation of this paper had little difference to official data and other researchers, so the estimation results was reliable. To further improve the emission inventory, local emission factors and technology change should be considered in the future.

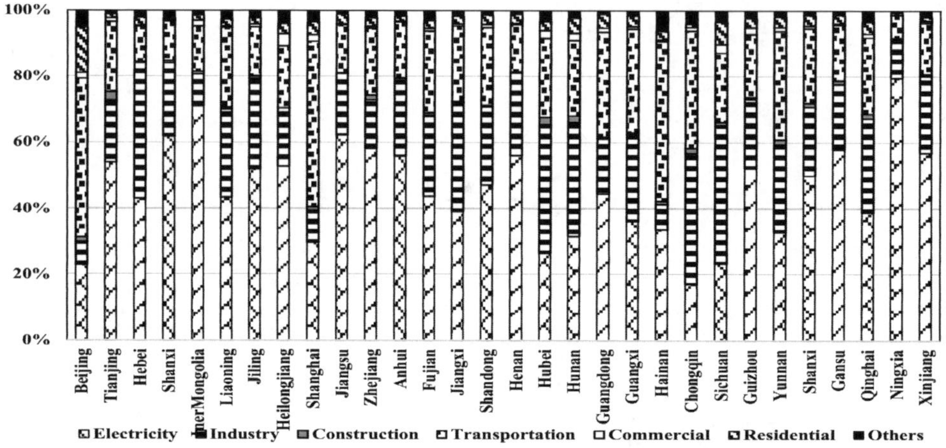

Figure 7. NOx speciation by provinces for the year 2012.

Acknowledgments

This work was supported by Humanities and Social Science Foundation of the Ministry of Education under Grant No. 13YJC790059, the Natural Science Foundation of Shandong (ZR2015DM014), the National Social Science Foundation of China under Grant No.11CJY039, and Humanities and Social Science Foundation of Shandong Provincial Department of Education under Grant No. J12WG59.

References

1. Wilfried Winiwarter, Zbigniew Klimont. The role of N-gases (N_2O, NOx, NH_3) in cost-effective strategies to reduce greenhouse gas emissions and air pollution in Europe. Current Opinion in Environmental Sustainability, **3**, 5 438–445 (2011)
2. Dincer I. Environmental impact of energy [J]. Energy Policy, **27**, 14 845–54 (1999)
3. Jun-Ichi Kurokawa, Keiya Yumimoto, Itushi Uno, et al. Adjoint inverse modeling of NOx emissions over eastern China using satellite observations of NO2 vertical column densities[J]. Atmospheric Environment, **43**, 4 1878–1887 (2009)
4. D.G. Streets, S.T. Waldhoff. Present and future emissions of air pollutants in China: SO_2, NO_x, and CO [J]. Atmospheric Environment, **34**, 2 363-374 (2000)
5. Tian He-Zhong, Hao Ji-Ming, Lu Yong-Qi, et al. Inventories and distribution characteristics of NO_x emissions in China [J]. China Environmental Science. **21**, 6 493-497 (2001)
6. Qing Lu, Junyu Zheng, Siqi Ye, et al. Emission trends and source characteristics of SO_2, NOx, PM_{10} and VOCs in the Pearl River Delta region from 2000 to 2009[J]. Atmospheric Environment, **76**, 1 11-20 (2013)
7. Yun Shi, Yin-Feng Xia, Bi-Hong Lu, et al. Emission inventory and trends of NOx for China, 2000–2020[J]. Journal of Zhejiang University-SCIENCE A, **15**, 6 454-464 (2014)
8. Chuying Zhang, Shuxiao Wang, Jia Xing, et al. Current status and future projections of NOx emissions from energy related industries in China [J]. Acta Scientiae Circumstantiae, **28**, 12 2470-2479 (2008)
9. Wang Wenxing, Wang Wei, Zhang Wanhua, et al. geographical distribution of SO2 and NOx emission intensities and trends in China. China environmental Science, **16**, 3 161-168 (1996)

MATEC Web of Conferences **64**, 06002 (2016) DOI: 10.1051/matecconf/20166406002
ICCMP 2016

Energetic Efficiency of the Vegetable Waste Used as Substrate for Biogas Production

Damian Janczak[1], Kamil Kozłowski[1],Zbyszek Zbytek[2], Marta Cieślik[3], Artur Bugała[4], and Wojciech Czekała[1]

[1]*Institute of Biosystems Engineering, Poznan University of Life Sciences, ul. Wojska Polskiego 28 60-637 Poznan, Poland*
[2]*Industrial Institute of Agricultural Engineering, ul. Starolecka 31 60-963 Poznan, Poland*
[3]*Department of Biotechnology and Food Microbiology, Poznan University of Life Sciences, Wojska Polskiego 48, 60-637 Poznan, Poland*
[4]*Institute of Electrical Engineering and Electronics, Poznan University of Technology, ul. Piotrowo 3a, 60-965 Poznan, Poland*

Abstract. Different vegetable waste materials (beetroot leaves, turnip leaves, yellow beans, onion waste), commonly found in Central Poland were tested for estimation of energy efficiency while methane fermentation process. The results showed that the biggest green mass of vegetable waste harvested from field was related to the highest water content. In contrary, the higher methane production from 1 Mg of tested biowaste was obtained in case of the materials with the highest dry mass content (yellow beans and onion waste, 28.03 and 44.33 m^3 Mg^{-1}). Although the onion waste was the lowest fermented substrate, however the energy yield for this material is visibly the highest (4435 kWh ha^{-1}) among all tested ones.

1 Introduction

Growing demand for electricity and heat as a result of economic development and increasing environmental pollution due to the use of fossil fuels (mainly coal) forced the mankind to look for alternative energy carriers. Hence, in recent years, the use of environment-friendly technologies i.e. renewable sources, had a significantly growing interest [1-3]. Implementation of renewable energy sources (RES) in recent years can be also related to a rational waste management and generally - increasing emphasis on environmental protection [4-5]. The most important advantage of renewable energy is the fact that RES are inexhaustible and renewed naturally.

One of the RES type is the energy used from biomass. Besides the pellets, briquetting, pyrolysis and direct combustion, the biomass is used as a substrate for biogas plants [6-7]. Those installations can produce the electricity and heat, and moreover allow to utilize the waste materials from crop production [8].

The biogas market in Europe is growing rapidly, wherein the essential substrates for installations are maize silage and animal faeces [9-10]. There have been observed new efforts and attempts to use the alternative substrates, including technologically advanced materials i.e. algae [6-11]. In Poland, with low support for each produced energy unit - MWe el, building the biogas plant installation with technology based on substrates from dedicated crops is highly unprofitable. Production of 1 kg of maize silage costs about 120-150 PLN, which is too high price, taking into account the fact that in Poland, the average price for produced electricity is approx. 3 times lower than in Germany. The advantage of a biogas plant is that it can use plant waste materials or waste from fruit and vegetable processing [12]. This fact may cause the development of the biogas market in Poland, since there are many regions producing millions of tons of organic waste from agriculture and agri-food industry. It should be noted that after joining the EU in 2004, Poland received large financial resources for investments in the both agricultural and food processing sector. As a result, there were implemented very modern solutions promoting the efficiency of agricultural production, food processing, even including those based on the use of artificial intelligence techniques [13-15], and image analysis [16-20]. Moreover, for environmental friendly processing of bio-waste many actions have been taken on a large scale. The fastest growing technology was the bio-waste composting [21-26]. However, it should be highlighted that methane fermentation of bio-waste ensures effective processing and also enables efficient production of biogas and in further consequence - electricity and heat [27]. In order to make a biogas plant a profitable investment, already planning the whole construction, a suitable substrate should be selected, which will be a year-round feedstock for biogas plants [28]. The most common substrate used in biogas plants in Europe is maize silage [29]. Promising substrates for biogas production are vegetable waste materials generated during the production and processing of field vegetables, such as beetroot leaves, turnip greens, yellow beans and husk onion. Like so far there are no biogas plants in Poland using waste generated during the production of field vegetables.

This will be certainly a good alternative for current management of these waste materials, which are distributed across the fields for fertilizing purposes - special permit required. Another method to manage those waste is composting process [25,30] however in this case – except the obtained fertilizer – the economy effect of the technology is negative. Taking into account the "waste-to-energy" idea, it can be stated that use of vegetable waste for energy production is a good direction for management of the waste of this kind.

2 Aim and the range of research

The aim of this study was to examine the biogas productivity of waste from typical agricultural vegetables as well as to analyze the energetic potential of those biowaste materials in case of use as a substrate in biogas plant. Because of the low support for electric energy production from RES in Poland, the main possibility to keep the positive economic balance for biogas plants is to avoid the usage of typical but expensive plants (i.e. maize or rye silage) and look for cheaper substrates like agricultural residues. The following research objectives have been formulated:

to analyze the potential of biogas production from different biowaste based on vegetable waste materials;

to calculate the energy value of the vegetable waste used in biogas production process.

3 Materials and methods of research

The analyzed materials were collected from the agricultural farms specialized in vegetable production. These farms are placed in south of Wielkopolska Region (Central Poland) which is the center of Polish vegetable production. The amount of vegetable waste materials in this area produced by farms and agro-food industry exceeds yearly over 1 million Mg, which directly shows the importance of this problem. Actually, this situation is mainly solved by spreading the biowaste on the fields which is often forbidden by Polish legislation.

The following vegetable waste materials have been tested:

- beetroot leaves,
- turnip leaves,
- yellow beans,
- onion waste (including rinds).

The research tests were made in the Laboratory of Ecotechnologies (which is the biggest Polish biogas laboratory), placed in the Institute of Biosystems Engineering, at the Poznan University of Life Sciences (PULS). The methodology of measurement and the biogas production calculations were made in accordance with the following norms: DIN 38 414/S8 and VDI 4630. The physical and chemical analyzes were made according to the Polish standard systems: dry matter content under Polish standard norm PN-75C-04616/01 (drying by 24 h in 105°C), organic dry matter according to the norm PN-Z-15011-3 (combustion of the samples in 525°C by 3 hours).

The fermentation set-up consisted of 21 biofermentors. Each individual biofermentor (made from glass) had a volume of 2 dm^3. The special design of the biofermentor as well as the whole position has kept the experiments under anaerobic conditions. The process was carried out under mesophilic conditions at 39°C +/-1°C. The scheme of experimental set-up for biogas fermentation is shown in Figure 1.

The produced biogas in each biofermentor chamber was transported via teflon pipe to the gas reservoir. These reservoirs were made from plexiglass as an inverted cylinder immersed in water. Between the water and gas areas, there was a liquid barrier preventing the dissolution of CO_2 in the water. According to the DIN norm, all samples were tested in triplicate.

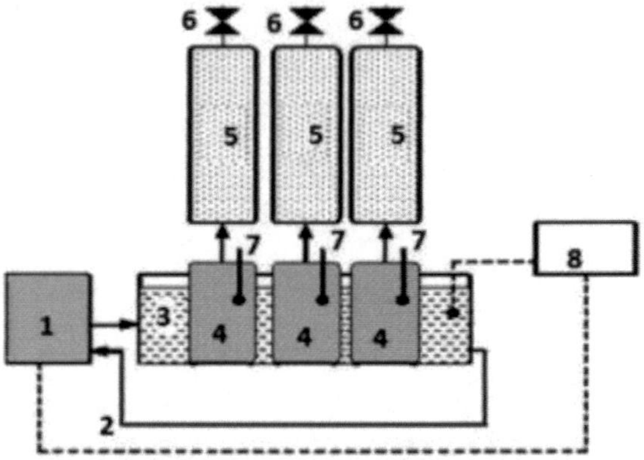

Figure 1. Scheme of biofermenter for biogas production research (3-chamber section). 1: Water heater with temperature regulator, 2: insulated conductors of calefaction liquid, 3: water coat with temperature 39oC, 4: biofermentor with charge capacity 2 dm3, 5: biogas reservoir, 6: cutting off valves, 7: sampling tubes, 8: recording central station.

4 Results

4.1 The potential of biogas production

Before taking the samples for analyzing biogas efficiency, the estimation of vegetable waste yield has been made. The results are shown in Table 1.

Table 1. The parameters of the tested material.

Type of biowaste	Harvested mass [$Mg*ha^{-1}$]
Beetroot leaves	25
Turnip leaves	20
Yellow beans	10
Onion waste	14

The presented results show that the biggest productivity is in case of beetroot leaves (25 Mg ha^{-1}). However, the dry mass (T.S.) concentration as well as organic matter content (V.S.) is quite low for this biowaste (Table 2). This suggests that energy efficiency of this material can be lower comparing to the other kinds of biowaste – with higher level of dry mass.

The initial analysis of all tested biowaste is shown in Table 2.

Table 2. The basic parameters of tested biowaste: Dry mass (T.S.), organic dry mass (V.S.) and pH.

Type of biowaste	T.S. [%]	V.S. [%]	pH [-]
Beetroot leaves	9.68	71.30	6.61
Turnip leaves	8.93	82.00	6.08
Yellow beans	11.77	93.97	5.45
Onion waste	28.13	72.81	8.49

Except the onion waste, all materials were characterized by low content of dry matter. In all cases, the dry matter concentration was significantly lower comparing to maize straw – the most popular substrate for biogas production in Central Europe. The concentration of organic matter was high in all cases, which was favourable for methane fermentation.

Table 3. The biogas and biomethane production from fresh mass of analysed biowaste.

Type of biowaste	CH_4 content [%]	CH_4 yield [m^3 Mg^{-1}]	Biogas yield [m^3 Mg^{-1}]
Beetroot leaves	52.61	17.63	33.51
Turnip leaves	46.10	23.90	51.84
Yellow beans	46.89	28.03	59.79
Onion waste	51.34	44.33	86.36

As it was expected, because of higher content of dry mass, the highest biogas and biomethane productivity was found in case of onion waste. Even the highest methane concentration was observed for beetroot leaves (52.61%), the total production of biomethane was the lowest (17.63 m^3 Mg^{-1}).

It should be also underlined, that the dynamics of the biogas production was different for all tested biowaste materials (Figure 2).

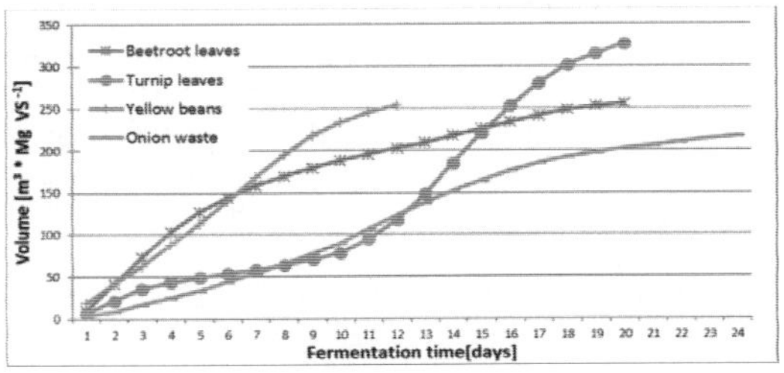

Figure 2. Cumulative production of methane from 1 Mg of VS of tested materials.

The results showed on Fig. 2 proved considerable differences in fermentation time (HRT) for tested materials. Yellow beans waste has the shortest HRT period which is favourable for intensive exploitation of biogas plant installation. Onion waste is the worst material (among tested) in terms of intensity of biogas production and duration of fermentation.

4.2 Energetic analysis

The comparison of tested biowaste energy yield based on biomass harvested from the fields and biogas production efficiency is shown in Table 4.

Table 4. Energy yield of analysed biowaste.

Sample	CH_4 yield [m^3 Mg^{-1}]	EEP* [kWh Mg^{-1}]	Mass yield [Mg*ha^{-1}]	Energy yield [kWh ha^{-1}]
Beetroot leaves	33.51	0.123	25	3073
Turnip leaves	51.84	0.19	20	3803
Yellow beans	59.79	0.219	10	2193
Onion waste	86.36	0.317	14	4435

* Electric Energy Potential, energetic value of 1 Mg of fresh mass, calculated as CH_4 volume multiplied by energetic value of CH_4 (9.17 kWh m^{-3}) and electric energy coefficient of cogeneration unit (40%)

The comparison of tested biowaste shows that the best energy yield was noted in case of onion waste (4435 kWh ha^{-1}) and the lowest for yellow beans waste (2193). Those values however are visibly lower comparing to the energy yield of typical agricultural substrate in Europe – maize silage (16836 kWh ha^{-1}). From economical point of view, the analyzed substrates are offered for free, or for the cost of transport, comparing to maize silage with the high price 27-50 Euro/Mg. Thus, utilization of the aforementioned waste materials seems to be better from economic reasons, even the biogas productivity is lower.

5 Conclusions

The research experiments described in this paper led us to the following conclusions:

• According to the results shown in Tables 1 and 2, the biggest fresh mass of vegetable waste (beetroot leaves) harvested from field was related to the highest water content. However, the highest harvested dry matter amount was in case of onion waste.

• The higher methane production from 1 Mg of tested biowaste was obtained in the case of the materials with the highest dry mass content (yellow beans and onion waste, 28.03 and 44.33 m^3 Mg^{-1}).

• The onion waste was the lowest fermenting substrate, based on fresh mass calculation, however the energy yield for this material is visibly the highest (4435 kWh ha^{-1}) among all tested materials.

• The energy yield of the best tested material (onion waste) is almost 4 times lower than in case of maize silage (16836 kWh ha^{-1}), however high price of maize silage in Central Europe (27-50 Euro/Mg) makes this material less attractive without important subsidies for biogas production.

References

1. W. Czekała, K. Kozłowski, J. Dach, P. Boniecki, A. Lewicki, D. Janczak, K. Jóźwiakowski, T. Piechota, Energy Conversion from Biomass to Hydrogen and Methane. 4th International Conference on Materials Engineering for Advanced Technologies (ICMEAT) Location: London, England, pp. 654-657 (2015)
2. J.P. Maity, J. Bundschuh, C-Y. Chen, P. Bhattacharya, Microalgae for third generation biofuel

production, mitigation of greenhouse gas emissions and wastewater treatment: Present and future perspectives - A mini review. Energy **78**, 104–113 (2014)
3. J. Dach, K. Koszela, P. Boniecki, M. Zaborowicz, A. Lewicki, W. Czekała, J. Skwarcz, Q. Wei, H. Piekarska-Boniecka, I. Białobrzewski, The use of neural modeling to estimation the methane production from slurry fermentation processes. Renewable and Sustainable Energy Reviews **56,** 603–610 (2016)
4. D. Janczak, M. Marciniak, A. Lewicki, W. Czekała, K. Witaszek, P.C. Rodríguez Carmona, M. Cieślik, J. Dach, Bioreactor Internet System for Experimental Data Monitoring and Measurement, Procedia Technology **8**, 209–214 (2013)
5. K. Malińska, M. Zabochnicka-Świątek, J. Dach, Effects of biochar amendment on ammonia emission during composting of sewage sludge. Ecological Engineering **71**, 474-478 (2014)
6. A. Lewicki, J. Dach, D. Janczak, W. Czekala, The experimental macro photoreactor for microalgae production, Procedia Technology **8**, 622-627 (2013)
7. F. Dong, J. Lu, Using solar energy to enhance biogas production from livestock residue - A case study of the Tongren biogas engineering pig farm in South China. Energy **57**, 759-765 (2013)
8. Q. Xinshan, Z. Shuping, W. Yuzhi, W. Renqing, Advantages of the integrated pig-biogas-vegetable greenhouse system in North China. Ecological Engineering **24**, 177–185 (2005)
9. J. Dach, W. Czekała, P. Boniecki, A. Lewicki, T. Piechota, Specialised Internet Tool for Biogas Plant Modelling and Marked Analysing. Advanced Materials Research **909**, 305-310 (2014)
10. A. Lewicki, P. Boniecki, J. Dach, D. Janczak, W. Czekała, Z. Zbytek, M. Cieślik, Neural Estimation of Biomethane Efficiency from Maize Silage, Maize Straw and Cattle Slurry. 4th International Conference on Materials Engineering for Advanced Technologies (ICMEAT) Location: London, (2015)
11. S. Cerbin, K. Nowakowski, J. Dach, K. Pilarski, P. Boniecki, J. Przybyl., A. Lewicki, Possibilities of neural image analysis implementation in monitoring of microalgae production as a substrate for biogas plant. 4th International Conference On Digital Image Processing (ICDIP 2012). Proceedings of SPIE **8334** (2012)
12. S. Berlian, Sukandar, D. P. Seno, Biogas recovery from anaerobic digestion process of mixed fruit - vegetable wastes. International Conference on Sustainable Energy Engineering and Application. Energy Procedia **32**, 176–182 (2013)
13. P. Boniecki, K. Nowakowski, R. Tomczak, S. Kujawa, H. Piekarska-Boniecka, The application of the Kohonen neural network in the non-parametric quality-based classification of tomatoes. 4th International Conference on Digital Image Processing. Kuala Lumpur (Malaysia), proceedings of SPIE, **8334**, (2012)
14. W. Czekała, J. Dach, A. Ludwiczak, A. Przybylak, P. Boniecki, K. Koszela, M. Zaborowicz, K. Przybył, D. Wojcieszak, The use of image analysis to investigate C:N ratio in the mixture of chicken manure and straw. Proc. SPIE. 9631, Seventh International Conference on Digital Image Processing (ICDIP 2015), **963117** (2015)
15. P. Boniecki, H. Piekarska-Boniecka, K. Koszela, M. Zaborowicz, K. Przybył, D. Wojcieszak, Z. Zbytek, A. Ludwiczak, A. Przybylak, A. Lewicki, Neural classifier in the estimation process of maturity of selected varieties of apples. Proc. SPIE. 9631, Seventh International Conference on Digital Image Processing (ICDIP 2015), **963115** (2015)
16. S. Kujawa, R.J. Tomczak, T. Kluza, J. Weres, P. Boniecki, A stand for the image acquisition of composted material based on the sewage sludge. 4th International Conference on Digital Image Processing. Kuala Lumpur (Malaysia), proceedings of SPIE, **8334** (2012)
17. K. Nowakowski, P. Boniecki, R.J. Tomczak, S. Kujawa, B. Raba, Identification of malting barley varieties using computer image analysis and artificial neural networks. 4th International Conference on Digital Image Processing. Kuala Lumpur (Malaysia), proceedings of SPIE, **8334** (2012)
18. D. Janczak, A. Lewicki, R. Mazur, P. Boniecki, J. Dach, J. Przybyl, M. Pawlak, K. Pilarski, W. Czekala, The Selected Examples of the Application of Computer Image Analysis in the Assessment of Environmental Quality. 5th International Conference on Digital Image Processing. Beijing (China), proceedings of SPIE, **8878** (2013)
19. K. Przybył, K. Koszela, P. Boniecki, W. Mueller, B. Raba, A. Lewicki, Organoleptic Damage Classification of Potatoes with the Use of Image Analysis in Production Process, Proc. SPIE 9159, Sixth International Conference on Digital Image Processing (ICDIP 2014) (2014).
20. B. Raba, K. Nowakowski, A. Lewicki, K. Przybył, M. Zaborowicz, K. Koszela, P. Boniecki, The Non-touching Method of the Malting Barley Quality Evaluation. Proc, SPIE 9159, Sixth International Conference on Digital Image Processing (ICDIP 2014) (2014)
21. A. Wolna-Maruwka, J. Dach, Effect of Type and Proportion of Different Structure-Creating Additions on the Inactivation Rate of Pathogenic Bacteria in Sewage Sludge Composting in a Cybernetic Bioreactor. Archives of Environmental Protection. **35,** 3 87-100 (2009)
22. K. Malińska, M. Zabochnicka-Świątek, Selection of bulking agents for composting of sewage sludge. Environment Protection Engineering **39**, 2 91-103 (2013)
23. 23. A. Piotrowska-Cyplik, Ł. Chrzanowski, P. Cyplik, J. Dach, A. Olejnik, J. Staninska, J. Czarny, A. Lewicki, R. Marecik, J. Powierska-Czarny, Composting of oiled bleaching earth: Fatty acids degradation, phytotoxicity and mutagenicity changes. International Biodeterioration & Biodegradation. **78** 49-57 (2013)
24. K. Koszela, J. Otrząsek, M. Zaborowicz, P. Boniecki, W. Mueller, B. Raba, A. Lewicki, K. Przybył,

Quality Assessment of Microwave-vacuum Dried Material with the Use of Computer Image Analysis and Neural Model, Proc. SPIE 9159, Sixth International Conference on Digital Image Processing (ICDIP 2014) (2014)
25. A. Lewicki, J. Dach, P. Boniecki, W. Czekała, K. Witaszek, The Control of Air Humidity and Temperature in Relationship with a Biowaste Composting Process. Advanced Materials Research **909** 455-462 (2014)
26. I. Białobrzewski, M. Mikš-Krajnik, J. Dach, M. Markowski., W. Czekała, K. Głuchowska, Model of the sewage sludge-straw composting process integrating different heat generation capacities of mesophilic and thermophilic microorganisms. Waste Management **43** 72-83 (2015)
27. J. Dach, P. Boniecki., J. Przybył, D. Janczak, A. Lewicki, W. Czekała, K. Witaszek, P.C. Rodríguez Carmona, M. Cieślik, Energetic efficiency analysis of the agricultural biogas plant in 250 kWe experimental installation. Energy **69** 34-38 (2014)
28. M. Lebuhn, A. Hanreich, M. Klocke, A. Schlüter, C. Bauer, C. Pérez, Towards molecular biomarkers for biogas production from lignocellulose-rich substrates. Anaerobe, **29** 10-21 (2014)
29. E. Bruni, A.P. Jensen, E.S. Pedersen, I. Angelidaki, Anaerobic digestion of maize focusing on variety, harvest time and pretreatment. Applied Energy. **87** 2212–2217 (2010)
30. J. Starzyk, W. Czekała, The Influence of Admixtures Accelerating the Pine Bark Composting Process on Variation in the Bacteriological State of Composts. Archives of Environmental Protection **40**, 4, 125–135 (2014)

Tensioned Fabric Structures with Surface in the Form of Chen-Gackstatter

Hooi Min Yee, and Mohd Nasir Abdul Hadi

Universiti Teknologi MARA, 13500 Permatang Pauh, Pulau Pinang, Malaysia

Abstract. Form-finding has to be carried out for tensioned fabric structure in order to determine the initial equilibrium shape under prescribed support condition and prestress pattern. Tensioned fabric structures are normally designed to be in the form of equal tensioned surface. Tensioned fabric structure is highly suited to be used for realizing surfaces of complex or new forms. However, research study on a new form as a tensioned fabric structure has not attracted much attention. Another source of inspiration minimal surface which could be adopted as form for tensioned fabric structure is very crucial. The aim of this study is to propose initial equilibrium shape of tensioned fabric structures in the form of Chen-Gackstatter. Computational form-finding using nonlinear analysis method is used to determine the Chen-Gackstatter form of uniformly stressed surfaces. A tensioned fabric structure must curve equally in opposite directions to give the resulting surface a three dimensional stability. In an anticlastic doubly curved surface, the sum of all positive and all negative curvatures is zero. This study provides an alternative choice for structural designer to consider the Chen-Gackstatter applied in tensioned fabric structures. The results on factors affecting initial equilibrium shape can serve as a reference for proper selection of surface parameter for achieving a structurally viable surface.

1 Introduction

Tensioned Fabric Structures (TFS) are the composition of tensioned fabric as structural members where the fabric patterns are joined together at seams. It is then tensioned through mechanicals means or cables to a rigid supporting system to provide roofing structure. Figure 1 shows the example of tensioned fabric structure.

TFS include a wide variety of systems that are distinguished by their reliance upon tensile only members to support load. TFS have been employed throughout recorded history as in rope bridges and tents. However, large permanent tension structures were generally a 19th century development in bridges and a 20th century development in buildings. Tensioned fabric structure is the best alternative to cover area with low cost. One of the greatest TFS benefits is strength to weight ratio.

According to [1], TFS are durable structures and are good in insulation, light transmission and fire protection. All of these properties will be depending on the types of coating material used such as polyvinylchloride (PVC), polytetrafluoroethylene (PTFE) and ethylenetetrafluorethylene (ETFE). In addition, the membrane of TFS is made out of fibres. Those fibres are commonly made from nylon, polyester, glass and aramids.

TFS are normally designed to be in the form of equal tensioned surface. Minimal surface such as classical minimal surfaces, Costa and Möbius strip or their variation have been studied as possible choice of surface form for TFS by [2]-[11].

Tensioned fabric structure is highly suited to be used for realizing surfaces of new forms. However, none of the new examples mentioned present any results on the Chen-Gackstatter as load carrying members. Understanding of the possible Chen-Gackstatter initial equilibrium shapes to be obtained will provide alternative shapes for designers to consider. Form-finding using nonlinear analysis method can be used for both the initial equilibrium problem and load analysis for a new form of Chen-Gackstatter. This study is aimed at proposing a Chen-Gackstatter for structural designer to consider the Chen-Gackstatter applied in TFS.

Figure 1. Denver International Airport (Redevelopment), 2014.

2 Generation of chen-gackstatter

The name of Chen-Gackstatter is the name of two mathematicians Chi Cheng Chen and Fritz Gackstatter.

This shape was found in 1982. According to [12] and [13] Chen-Gackstatter as appeared in the Figure 2 was the first complete orientable minimal surfaces of R3 and Chen-Gackstatter is derived from Enneper's minimal surface. It forms a double-indexed collection 〖M〗_(i j), where i ⩾0 and j⩾1. The addition of i makes the surface has topological genus equals to i.

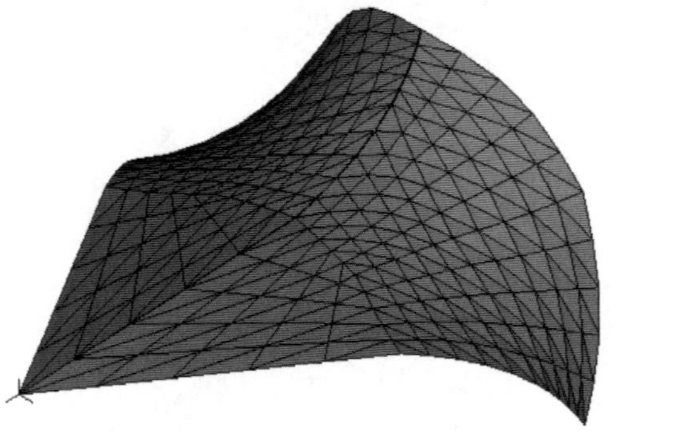

Figure 2. Chen-gackstatter minimal surface.

Figure 2 shows a model of Chen-Gackstatter. In this study, the XYZ coordinate of Chen-Gackstatter has been generated from Eq. (1). The software [14] has been used in this study.

$$X = \frac{4A^2}{3r^3}\cos(3\theta)$$

$$Y = -\frac{4A^2}{3r^3}\sin(3\theta)$$

$$Z = \frac{2A}{r^2}\cos(2\theta) \tag{1}$$

$$A = \sqrt{\frac{3\pi}{2g_2}}$$

$$g_2 = 60\sum\nolimits_{m,n=-\infty}^{\infty}\frac{1}{(m+ni)^4}$$

i = imaginary unit
r = small positive constant

$$0 \leqslant \theta \leqslant 2$$

g_2 = real and positive

3 Form-finding using nonlinear analysis method

The principle of proposed nonlinear analysis method by [2] is based on the large displacement finite element formulation used for analysis of structural behaviour under external loads. Since the method can be used for both the initial equilibrium problem and load analysis, the approach using nonlinear analysis is quite common. The basic equation used is expressed as follows:

$$({}_0^t\boldsymbol{K}_L + {}_0^t\boldsymbol{K}_G)\boldsymbol{u} = {}^{t+\Delta t}\boldsymbol{F} - {}_0^t\boldsymbol{f} \tag{2}$$

where ${}_0^t\boldsymbol{K}_L$ is linear strain incremental stiffness matrix, ${}_0^t\boldsymbol{K}_G$ is nonlinear strain incremental stiffness matrix, ${}_0^t\boldsymbol{f}$ is vector internal forces, ${}^{t+\Delta t}\boldsymbol{F}$ is load vector and $\boldsymbol{u}$ is vector of increment in displacement.

Nonlinear finite element analysis procedures for stress analysis of TFS have been used as basis for form-finding analysis in this study. As a first shape for the start of form-finding analysis procedure adopted in this study, initial guess shape is needed. The software [14] has been used for the purpose of model generation. For the generation of such initial guess shape, anticlastic feature is incorporated into the model in order to produce a better initial guess shape. Such anticlastic feature has been incorporated by means of specification of selected sag, Δ, relative to two suitably chosen points on the fixed boundary with span, L. Using this Δ/L sag ratio, geometry for approximates the surface of tensioned fabric structure is then generated. Such geometry is then meshed to produce initial guess shape with anticlastic feature.

The proposed computational strategy involves two stages of analysis in one cycle. The first stage (denoted as SF1) is an analysis which starts with an initial assumed shape in order to obtain an updated shape for initial equilibrium surface. This is then followed by the second stage of analysis (denoted as SS1) aimed at checking the convergence of updated shape obtained at the end of stage SF1. During stage SF1, elastic modulus E with very small values, are used. Warp and fill stresses, σ_W and σ_F are kept constant. In the second stage SS1, the actual tensioned fabric properties values are used. Resulting warp and fill stresses are checked at the end of the analysis against prescribed stresses. Iterative calculation has to be carried out in order to achieve convergence. The resultant shape at the end of iterative step n (SSn) is considered to be in the state of initial equilibrium under the prescribed warp and fill stresses and boundary condition if difference between the obtained and the prescribed tensioned fabric stresses relative to the prescribed stress satisfied the specified criteria. Such checking of difference in the obtained and prescribed stresses has been presented in the form of total stress deviation in warp and fill direction versus stress analysis stage.

In this study, three variable r=0.3, r=0.6 and r=0.9 of Chen-Gackstatter has been studied. This surface has been generated from Eq. (1) to create TFS model of Chen-Gackstatter. The number of nodes and triangular elements are 291 and 500, respectively for Chen-Gackstatter. The member pretension in warp and fill direction, denoted as σ_W and σ_F respectively, is 2000N/m. The shear stress is zero. Young's Modulus-Membrane, E.t, is 800000N/m, Shear modulus, Gt is 23529.41176N/m and Poisson's Ratio, $v_{xy} = v_{yx} =$ 0.7. In this study, form-finding using

 DOI: 10.1051/matecconf/20166407001

nonlinear analysis method proposed by [2] in Chen-Gackstatter has been carried out.

4 Result and discussion

4.1 Chen-Gackstatter, *r*=0.3

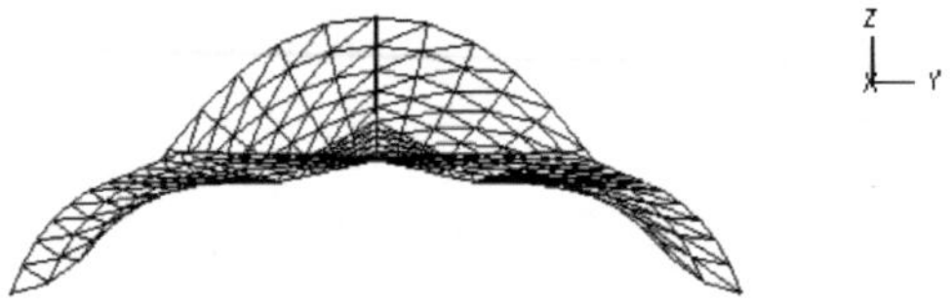

Front view

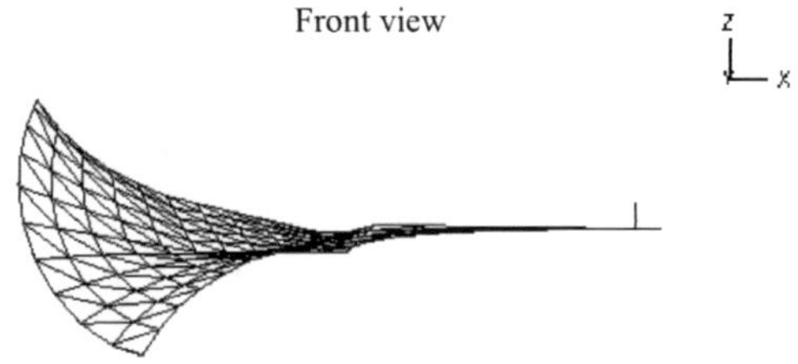

Side view

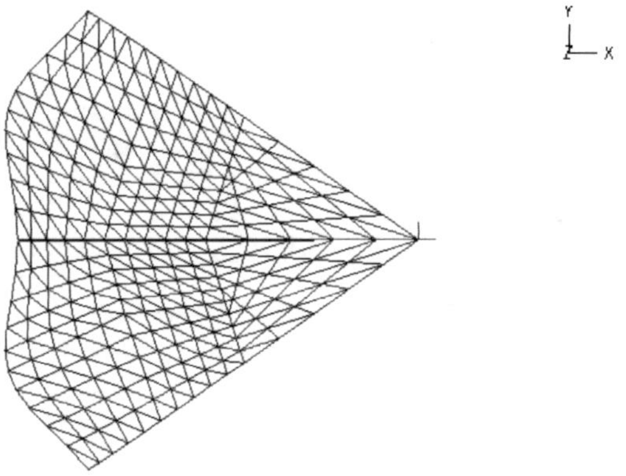

Top view

Figure 3. Initial GUESS shape of chen-gackstatter, *r*=0.3.

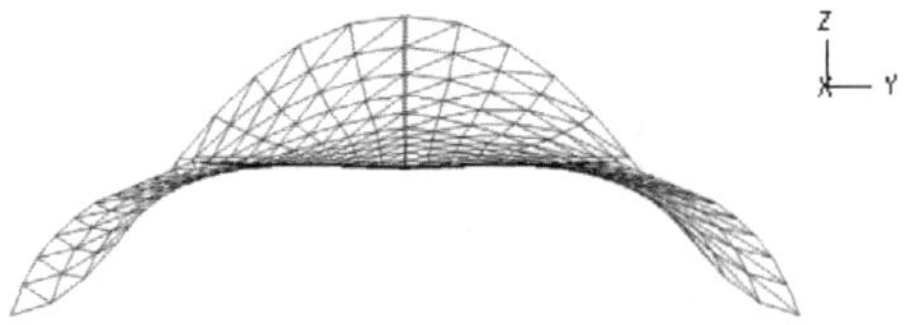

Front view

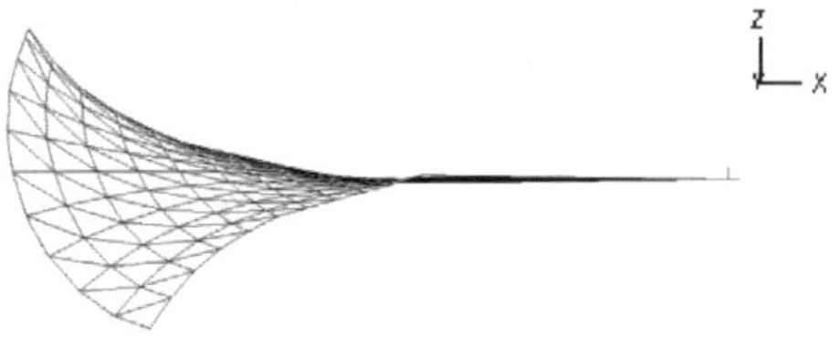

Side view

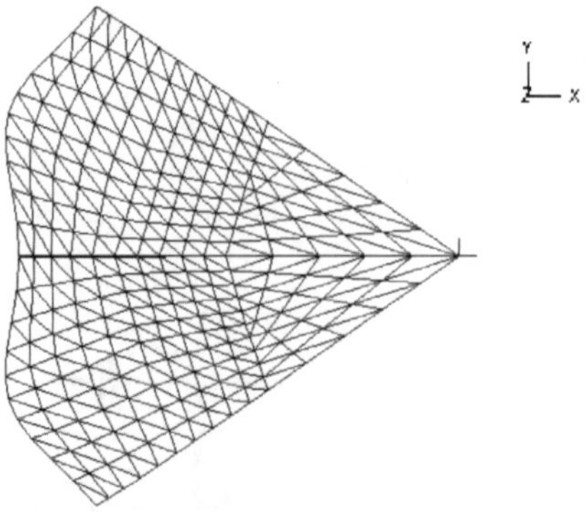

Top view

Figure 4. Converged shape of chen-gackstatter, *r*=0.3.

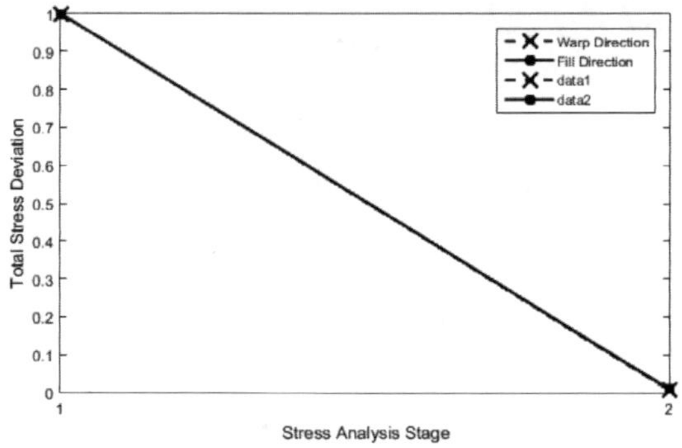

Figure 5. Convergent Curve of Chen-Gackstatter, *r*=0.3.

Figure 3 shows the initial guess shape of Chen-Gackstatter, r=0.3. Figure 4 shows the converged shape of the Chen-Gackstatter, r=0.3 after form-finding. Figure 5 also shows the convergent curve of the Chen-Gackstatter. From this result, the warp and fill direction for Chen-Gackstatter of r=0.3 is 0.006788 and 0.009175.The convergent curve shows that the total warp and fill stress deviation of Chen-Gackstatter, r=0.3<0.01.

4.2 Chen-gackstatter, *r*=0.6

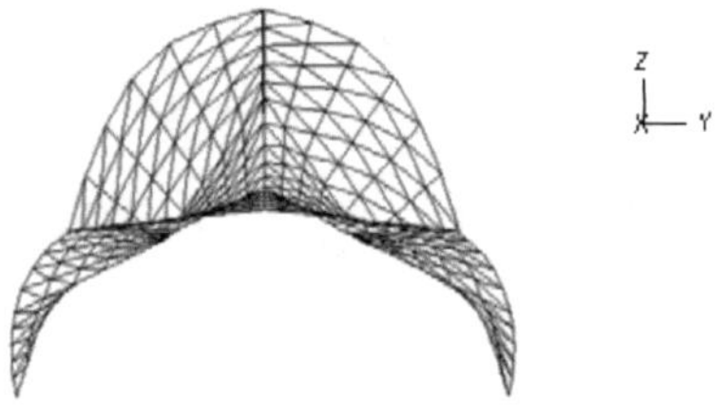

Front view

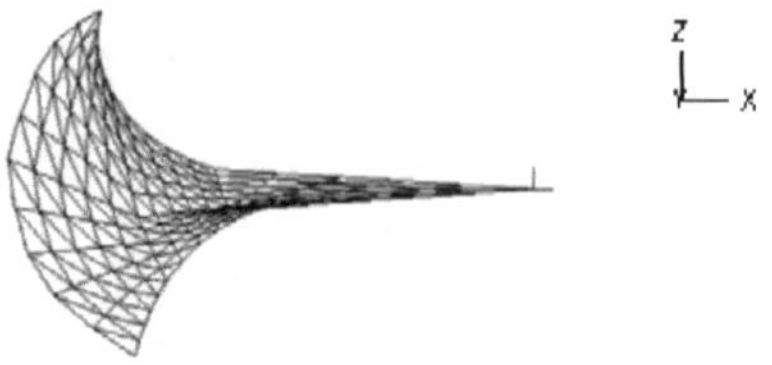

Side view

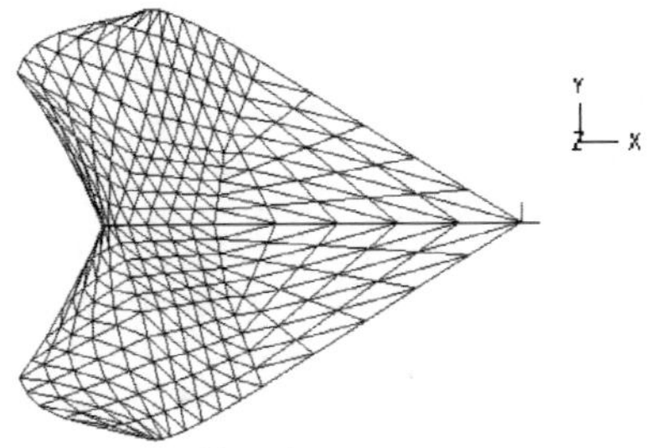

Top view

Figure 6. Initial guess shape of chen-gackstatter, *r*=0.6.

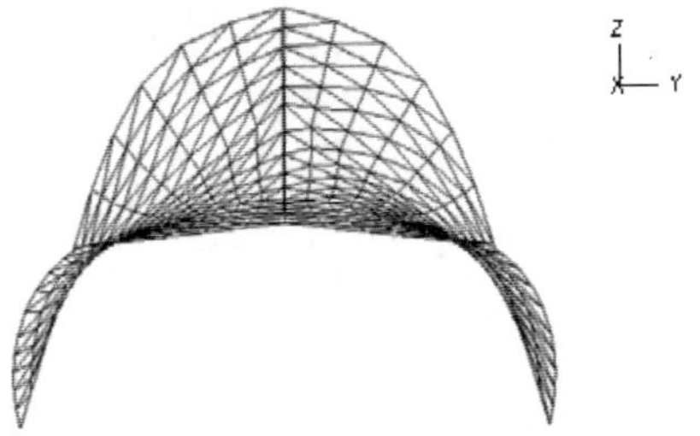

Front view

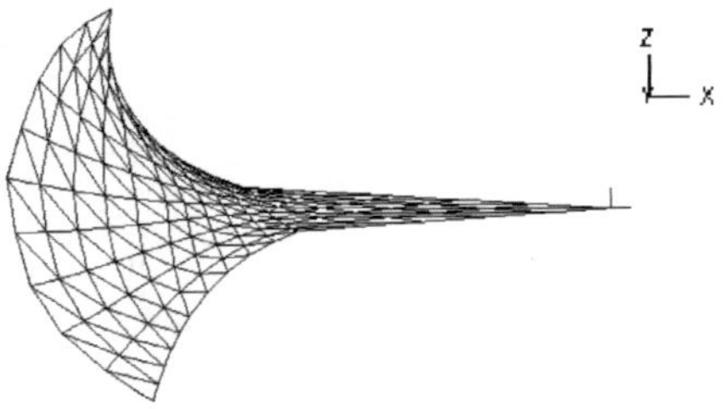

Side view

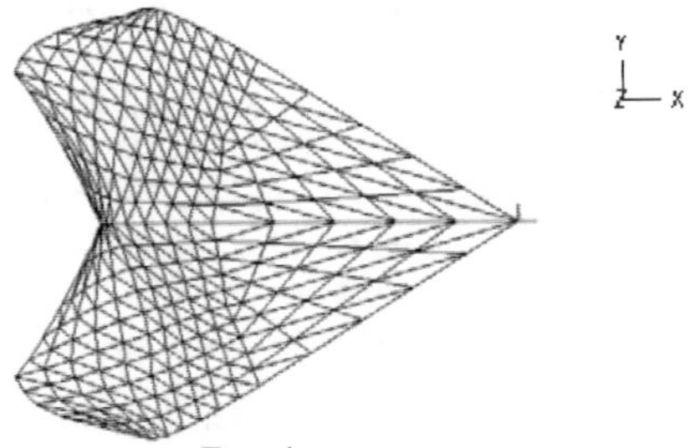

Top view

Figure 7. Converged shape of chen-gackstatter, *r*=0.6.

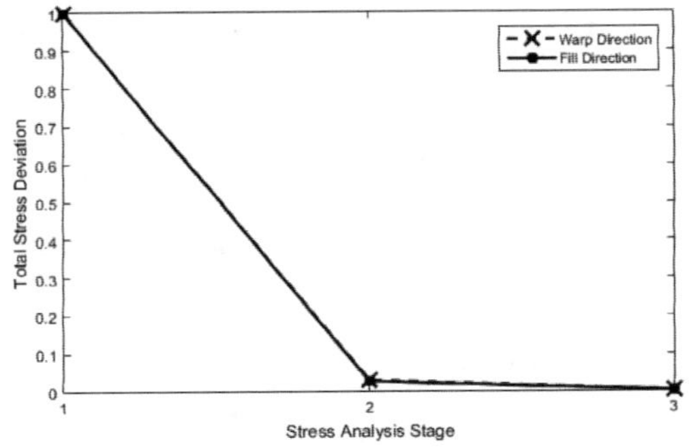

Figure 8. Convergent curve of chen-gackstatter, *r*=0.6

Figure 6 shows the initial guess shape of Chen-Gackstatter, r=0.6. Figure 7 shows the converged shape of the Chen-Gackstatter, r=0.6 after form-finding. Figure 8 also shows the convergent curve of the Chen-Gackstatter. The warp and fill direction for Chen-Gackstatter of r=0.6 is 0.001374 and 0.001211. The convergent curve shows that the total warp and fill stress deviation of Chen-Gackstatter, r=0.6<0.01.

4.3 Chen-gackstatter, *r*=0.9

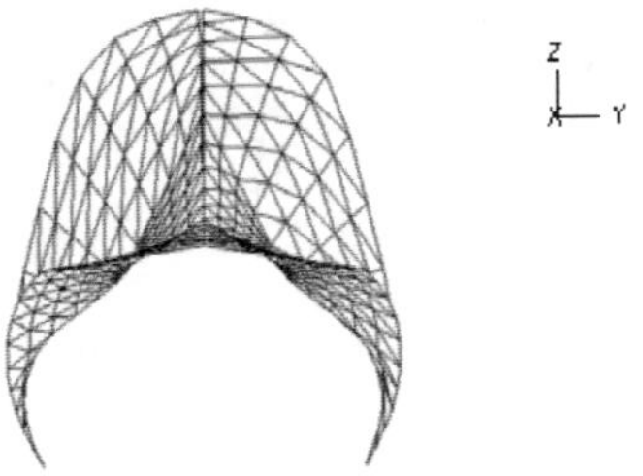

Front view

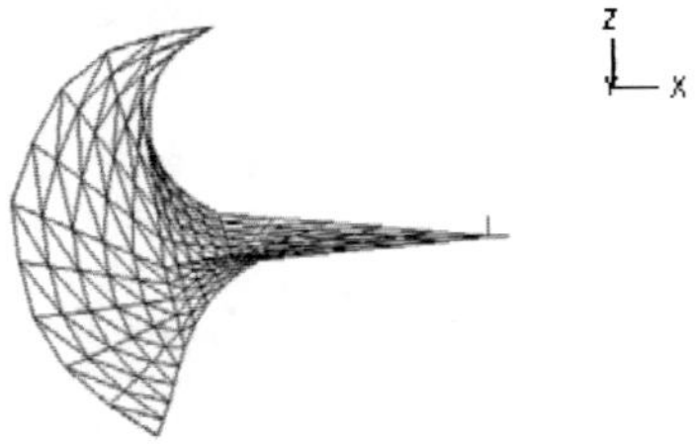

Side view

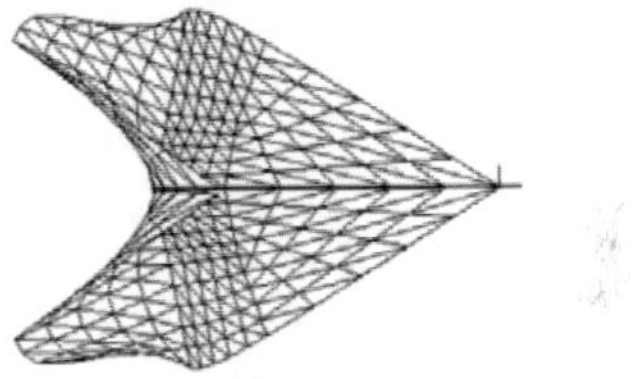

Top view

Figure 9. Initial guess shape of chen-gackstatter, *r*=0.9.

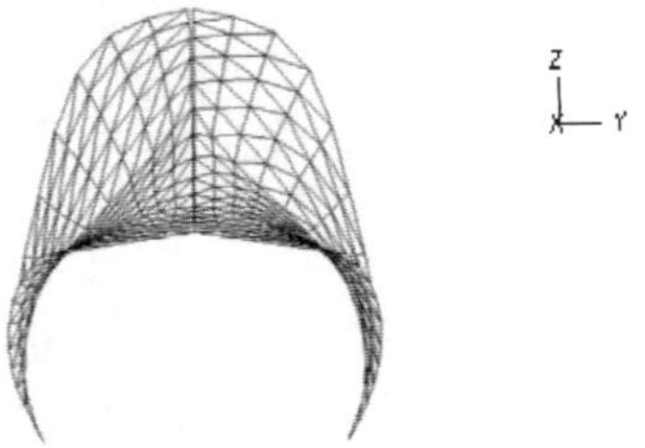

Front view

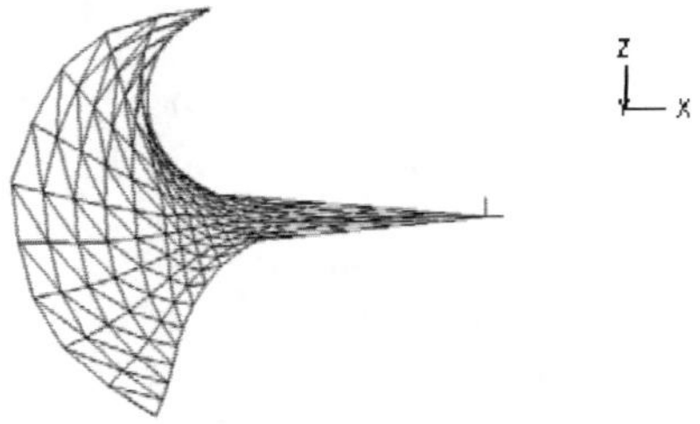

Side view

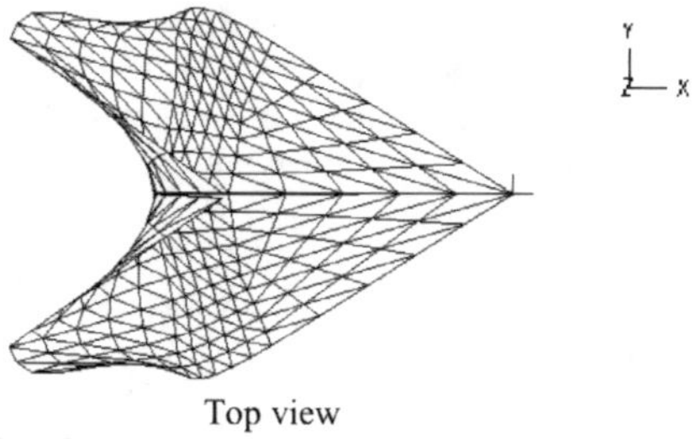

Figure 10. Converged shape of chen-gackstatter, *r*=0.9.

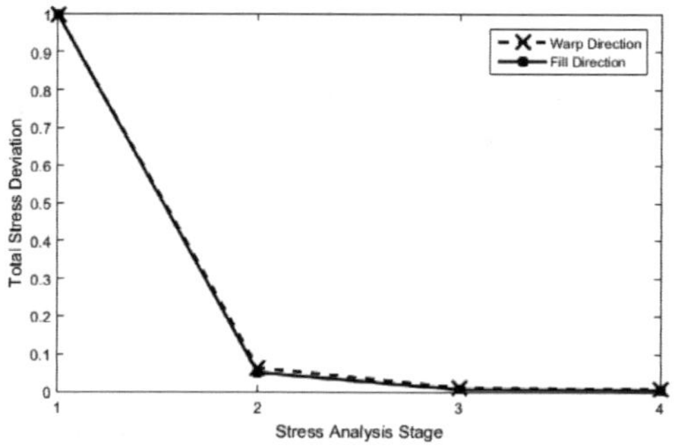

Figure 11. Convergent Curve of Chen-Gackstatter, *r*=0.9

Figure 9 shows the initial guess shape of Chen-Gackstatter, r=0.9. Figure 10 shows the converged shape of the Chen-Gackstatter, r=0.9 after form-finding. Figure 11 shows the convergent curve of the Chen-Gackstatter. The warp and fill direction for Chen-Gackstatter of r=0.9 is 0.008255 and 0.005117. The convergent curve shows that the total warp and fill stress deviation of Chen-Gackstatter, r=0.9<0.01.

5 Conclusion

Form Finding with Chen-Gackstatter with variable *r*=0.3, *r*=0.6 and *r*=0.9 has been carried out successfully using the procedure proposed by [2] which is based on nonlinear analysis method. The results from computational study show that TFS in the form of Chen-Gackstatter is a structurally viable surface form to be considered.

References

1. M.E. Son, The design and analysis of tension fabric structures, Massachusetts Institute of Technology, (2007)
2. H.M. Yee, A Computational Strategy for Form-Finding of Tensioned Fabric Structure using Nonlinear Analysis Method, Ph.D Dissertation, School of Civil Engineering, Universiti Sains Malaysia, Pulau Pinang, Malaysia (2011)
3. H.M. Yee, K.K. Choong, and J.Y. Kim, Form-Finding Analysis of Tensioned Fabric Structures Using Nonlinear Analysis Method, Advanced Materials Research, **243**, 249 1429-1434 (2011)
4. H.M. Yee and K.K. Choong, Form-Finding of Tensioned Fabric Structure in the Shape of Möbius Strip, Iranica Journal of Energy & Environment, **4** (2013)
5. H.M. Yee, K.K. Choong, and J.Y. Kim, Experimental Form-Finding for Möbius Strip and Ennerper Minimal Surfaces Using Soap Film Models, International Journal of Engineering Science and Innovative Technology, **2**, 328-335 (2013)
6. H.M. Yee, J.Y. Kim, and M.S. Mohd Noor, Tensioned Fabric Structures in Oval Form, Applied Mechanics and Materials, **405**, 408 1008-1011 (2013)
7. M.S. Mohd Noor, H.M. Yee, K.K. Choong, and A.H. Haslinda, Tensioned Membrane Structures in the Form of Egg Shape, Applied Mechanics and Materials, **405**, 408 989-992 (2013)
8. H.M. Yee and M.A. Samsudin, Development and Investigation of the Moebius Strip in Tensioned Membrane Structures, WSEAS Transactions on Environment and Development, **10**, 145-149 (2014)
9. H.H Yee, K.K Choong, and M. A. Hadi, Sustainable Development of Tensioned Fabric Green Structure in the Form of Enneper, International Journal of Materials, Mechanics and Manufacturing, **3**, 125-128 (2015)
10. H.M. Yee, A.H. Haslinda, and A.H. Nasir, Computer Investigation of Tensioned Fabric Structure in the Form of Enneper Minimal Surface, Applied Mechanics and Materials, 743-746 (2015)
11. H.M. Yee and M.A. Samsudin, Mathematical and Computational Analysis of Moebius Strip, International Journal of Mathematics and Computers in Simulation, **8**, 197-201 (2014)
12. E.C. Thayer, Higher-Genus Chen–Gackstatter Surfaces and The Weierstrass Representation for Surfaces of Infinite Genus, Experimental Mathematics, **4**, 19-39 (1995)
13. C.C. Chen and F. Gackstatter, Elliptische und hyperelliptische Funktionen und vollständige Minimalflächen vom Enneperschen Typ, Mathematische Annalen, **259**, 359-369 (1982)
14. K..J. Bathe, ADINA system, ADINA R&D Inc (2003)

DOI: 10.1051/matecconf/20166407002

Crushed Limestone Dust Valorization for Eco Friendly non Fired Wall Tiles

Ubolrat Wangrakdiskul, Kannika Benjaratthapong and Meatra Jongtrakulsiri

Department of Production Engineering, Faculty of Engineering, King Mongkut's University of Technology North Bangkok, Bangkok 10800, Thailand

Abstract. Utilizing crushed limestone dust; the by-product of crushing limestone plant for producing non-fired wall tiles is the objective of this research. The main raw materials of non- fired wall tiles consist of laterite soil, fluvial sand and Portland cement. The crushed limestone dust has been used as the substitute material of non-fired wall tiles for enhancing their bending strength. In this experiment, four groups of formulas have been constructed. Firstly, crushed limestone dust has been mixed into the basic formula for substitution of laterite soil. Secondly, crushed limestone dust has been used for substitution to river soil of the basic formula. Thirdly, it has been substituted to Portland cement in the basic formula. Finally, the best formulas of the first three groups has been selected and mixed up to make the new formulas. After mixing the raw materials of selected formula, forming specimens by uni-axial pressing at 100 bar has been performed. Then, they have been aged for 1 week at the room temperature. After that, testing the specimens' properties has been carried out. They are bending strength and water absorption which comparing with Thailand Industrial Standard (TIS) 614-2529. The experimental results show that adding the crushed limestone dust to the basic composition has an effect on the bending strength and water absorption. The compositions of the best formula are laterite soil 55%, fluvial sand 7.5%, Portland cement 22.5%, and crushed limestone dust 15%. It gives the bending strength by 2.32 MPa, water absorption 3.20%. The water absorption property can pass the TIS standard, but bending strength could not pass. The estimated cost of raw materials is 0.09065 THB/100 g.

1 Introduction

Crushed limestone is one of products for many construction uses including: Portland cement concrete aggregate, asphalt pavement (hot mix and warm mix) aggregate, asphalt surface treatments, road base, structural fill, railroad ballast, rip rap, and drainage and erosion control. In 2015, there are many limestone crushing plants in Thailand which are 314 plants. The total capacity is approximately 166 x 10^6 MT/year [1]. During the crushing process, by products such as crushed limestone dust having diameter less than 3 mm have been generated approximately 14.5% of total production. Its selling price is 260 THB/ton [2]. However they can be utilized as the substitute material for replacement in fluvial sand of concrete product. For used as value added material, limestone crushed dust has been studied by many researchers. Galetakis and Raka have mixed limestone dust with Portland cement for producing artificial stone. The results indicate that limestone can be used and give the acceptable compressive strength [3]. Utilization of limestone dust for brick making has been investigated in Egypt, finding of this research shows that 87% of limestone dust can be replaced on cement in the composition of bricks [4]. Fathi has studied about the effects of limestone crushed dust on the compressive strength and workability of concrete. This research indicated that limestone dust can be replaced on the natural sand in concrete composition with higher compressive strength [5].

Turgut and Halil have investigated the physical and mechanical properties of brick samples with wood sawdust wastes (WSW) and limestone powder wastes (LPW). The experiment shows that the WSW–LPW combination provides results which are of potential to be used in the production of lighter and economical new brick material [6]. Turgut has studied the utilization of different grades of wood sawdust wastes (WSW) and limestone powder wastes (LPW) for cement composites. The results indicated the high energy absorption capacity, reduced the unit weight dramatically and introduce a smoother surface compared to the current bricks in the market [7]. In 2008, Turgut investigated both physical and mechanical properties of samples containing limestone powder wastes (LPW) - waste glass powder (WGP) combinations for producing as new building brick material. An experimental approach to develop a new brick material including mainly LPW, a small quantity of Portland cement and WGP is presented. The WGP used in LPW remarkably improves the compressive strength, flexural strength, modulus of elasticity, abrasion resistance, F-T resistance, and thermal conductivity of LPW brick samples produced in this study [8]. Furthermore in 2013, Turgut also studied about the LPW and cotton waste material for replacement in artificial limestone composites. The energy conservation in buildings by using limestone powder waste and cotton waste composites having low thermal conductivity can reduce energy requirements [9]. Wangrakdiskul et al. have studied about non-fired wall tiles by replacement spent bleaching earth (SBE) in the basic formula having basic materials; laterite soil, fluvial sand, and Portland cement [10].

Most of previous researches mentioned above have focused on cement bricks production. Some research has used wastes from palm oil industry for non-fired wall tiles [10]. However, studying on utilizing limestone dust for non-fired wall tiles production is just not investigated. Therefore, this study aims to utilize crushed limestone dust in non-fired wall tiles. The experimental procedure will be described in the next section.

2 Experimental procedure

2.1 Materials

Basic materials used in this study consist of laterite soil, fluvial sand and Portland cement type I. The by-products material; crushed limestone dust from limestone crushing plant has been utilized for replacement on basic materials. Crushed limestone dust with diameter less than 3 mm has been used. The mixture proportions of basic formula for non-fired wall tiles have been provided from Wangrakdiskul, et al in 2008[11]. Mixing formulas of this experiment will be classified in following section.

2.2 Method of experimental research

All of materials as mentioned above have been prepared as the procedure illustrated in Fig. 1. It consists of 8 steps. In the 3rd step, formulas of varying the proportion of materials have been constructed which consist of 18 formulas (see Table 1, 2, and 3). The based formula No. 1_0 is obtained from Wangrakdiskul, et al [11]. Thirty pieces of specimens are performed for each formula. It could be explained in the following section.

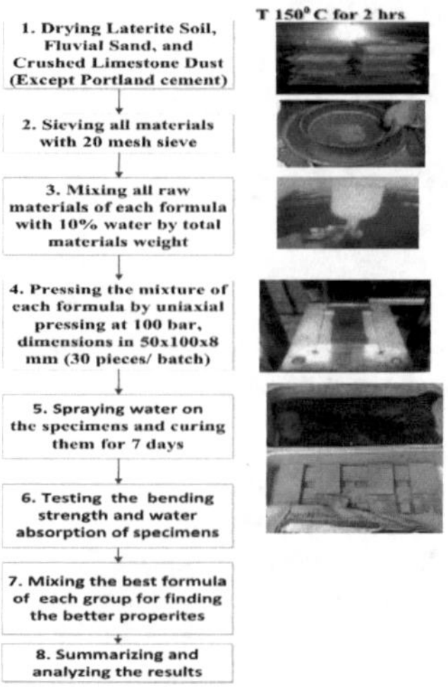

Figure 1. The experimental procedure.

Table 1. Composition formulas of varying %CLD by weight in laterite soil (Group 1).

No	%Fluvial Sand	%Portland Cement	%Laterite Soil	%Crushed Limestone Dust (CLD)
1_0	15	22.50	62.5	0
1_1	15	22.50	57.5	5
1_2	15	22.50	52.5	10
1_3	15	22.50	47.5	15
1_4	15	22.50	42.5	20
1_5	15	22.50	37.5	25
1_6	15	22.50	32.5	30
1_7	15	22.50	27.5	35
1_8	15	22.50	22.5	40
1_9	15	22.50	17.5	45
1_10	15	22.50	12.5	50
1_11	15	22.50	7.5	55
1_12	15	22.50	2.5	60

According to Table 1, 2, and 3, notice that formula No 1_0 is used as the basic formula of all groups for this experiment. Testing of bending strength and water absorption as indicated on step 6, the specimens have been performed in accordance with TIS 614-1986 and calculated the results by the formulas as expressed in Eq. 1 and Eq. 2.

$$B = 3\ PL \qquad (1)$$

where

B = Bending strength (MPa)

P = The maximum load applied to break a specimen (N)

L = Span length of beam supporting specimen (mm)

b = Width of specimen (mm)

d = Thickness of specimen (mm)

$$\%\text{Water absorption} = \underline{(Ww-Wd)}*100 \qquad (2)$$

where

Ww = Wet weight of specimen after placed in boiling water for five hours then soaked at room temperature water for 24 hours (g)

Wd = Dry weight of specimen (g)

After the properties of all formulas in three groups have been analysed and summarized by statistical analysis, the formulas of each group providing the high strength have been selected and then mixed with each other for producing the new formulas. The results will be described in the following section.

Table 2. Composition formulas of varying % CLD by weight in fluvial sand (Group 2).

No	%Fluvial Sand	%Portland Cement	%Laterite Soil	%Crushed Limestone Dust (CLD)
1_0	15	22.5	62.5	0
2_1	10	22.5	62.5	5
2_2	5	22.5	62.5	10
2_3	0	22.5	62.5	15

DOI: 10.1051/matecconf/20166407002

Table 3. Composition formulas of varying %CLD by weight in Portland cement (Group 3).

No	%Fluvial Sand	%Portland Cement	%Laterite Soil	%Crushed Limestone Dust (CLD)
1_0	15	22.5	62.5	0
3_1	15	17.5	62.5	5
3_2	15	12.5	62.5	10
3_3	15	7.5	62.5	15

3 Results and discussion

3.1 Mechanical and Physical Properties of group 1, 2, 3 and 4.

The properties of each formula have been classified into three groups in accordance with the group's formula; group 1, 2 and 3. They have been indicated in figure 2, 3 and 4.

According to fig 2, the results of replacement crushed limestone dust (CLD) to laterite soil have been illustrated. They can be summarized that the effect of CLD can enhance the bending strength. Obviously, the best formula that provides the highest bending strength is No.1_3. It consists of 15% of CLD. At the same time, %water absorption of this group is varying from 3.57-5.87. The best formula for lowest %water absorption is also No.1_3.

The results of replacement crushed limestone dust (CLD) to fluvial sand have been expressed in fig 3. The effects of CLD in fluvial sand can enhance the bending strength. The best formula that provides the highest bending strength is No.2_3. It consists of 15% of CLD, similarly the results of group 1. However, %water absorption of this group is higher than that of the basic formula No.1_0.

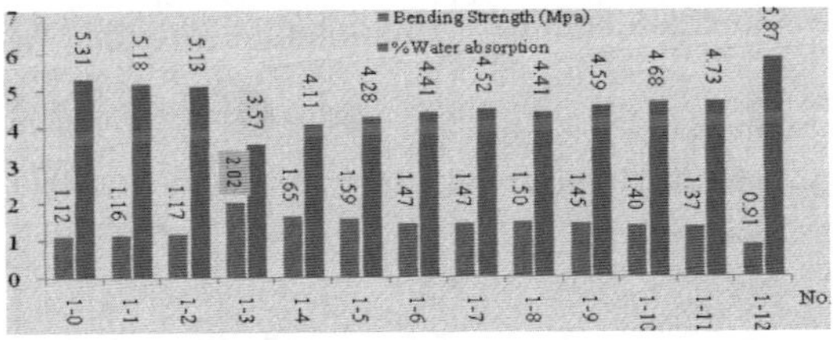

Figure 2. Bending strength and water absorption of group 1.

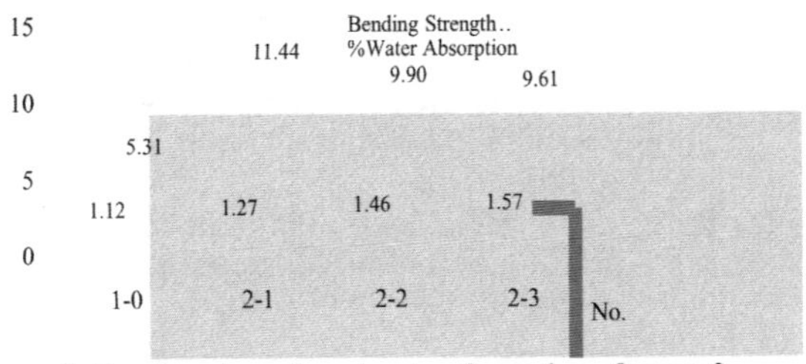

Figure 3. Bending strength and water absorption of group 2.

The effects of replacement CLD in Portland cement have been expressed in fig 4. As the increased of CLD of each formula, it gives the decreased of bending strength and increased of water absorption. The best formula of this group is No.3_1 which consists of 5%CLD. However, when comparing with the results of other groups, this group has not been selected for further improvement. Due to, the composition of CLD of the best formula is only 5%.

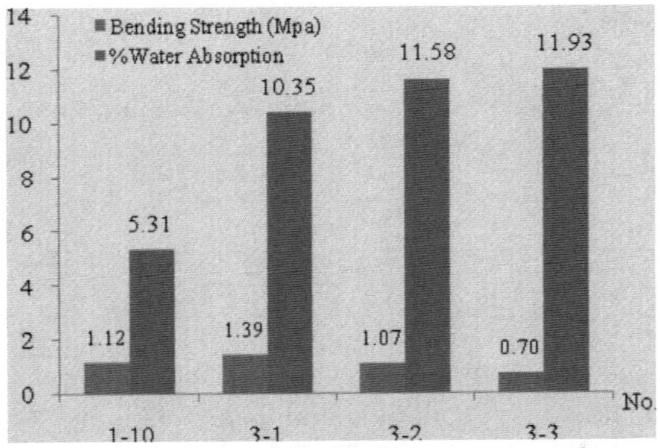

Figure 4. Bending strength and water absorption of group 3.

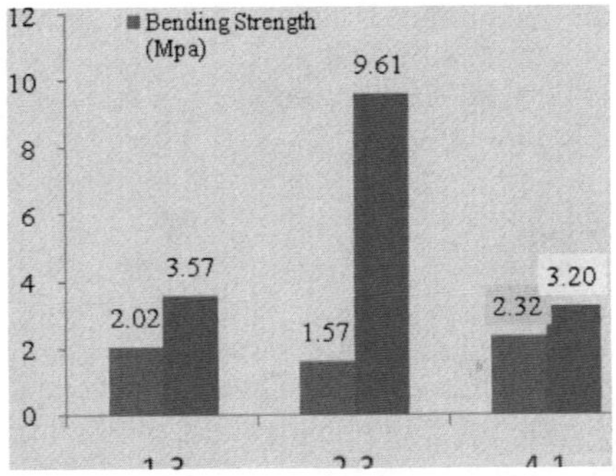

Figure 5. Bending strength and water absorption of group 4 mixing of the best formula from group 1 and 2.

Table 4. Composition formulas of No.4_1 the average composition of each material.

No	%Fluvial Sand	%Portland Cement	%Laterite Soil	%Crushed Limestone Dust (CLD)
1_3	15	22.5	47.5	15
2_3	0	22.5	62.5	15
4_1	7.5	22.5	55	15

Considering results of all three groups, the formula No.1_3 of group 1 and No.2_3 of group 2 have been selected for further investigated. Their compositions have been mixed to produce the new formula so called No.4_1. Its composition has been illustrated in table 4.

Bending strength and %water absorption of formula No.4_1 has been shown in fig 5. It can give the favorable properties, because the bending strength is higher than that of No.1_3 and No.2_3. Furthermore, its %water absorption is lower than that both of them.

4 Statistical analysis

For approving the results of this study, the experimental design has been conducted [12]. One way ANOVA has been employed for analyzing the effect of raw material replacement on the properties of each formula. P-values of hypothesis testing with a 0.05 significance level of the experiment are illustrated in Table 5.

Table 5. One way ANOVA test with a 0.05 significance level of varying raw materials on physical properties of each formula group.

Group	Bending Strength	%Water absorption
1	0.000	0.000
2	0.003	0.037
3	0.000	0.006
4	0.000	0.000

According to table 5, p-values of bending strength and water absorption for all experimental groups are less than 0.05. This means that replacement of CLD waste into main raw materials has an effect on bending strength and water absorption. However, the appropriate composition has been selected by considering on high bending strength and low %water absorption. In this study, formula No.4_1 is the most appropriate composition for more investigation.

5 Conclusion

As mentioned in the previous section, the best formula is No.4_1 with the highest bending strength and lowest water absorption. It can be concluded that CLD waste can fill up between the particles of laterite soil and of fluvial sand. Therefore, the water absorption is less than that of basic formula. Materials costs of all materials per 1000 g. and the best formula are 0.09065 THB ($0.00269 USD) per 100 g. of the test specimen which are shown in table 6 and 7, respectively.

Table 6. Costs of materials used for this study.

Raw Material	THB/ 1000 g
Fluvial Sand	0.235
Portland Cement	3
Laterite Soil	0.318
CLD	0.26

Table 7. Materials costs of the best formula (No.4_1).

Raw Material	% By weight	THB/ 100 g (piece)
Fluvial Sand	7.5	0.00176
Portland Cement	22.5	0.06750
Laterite Soil	55	0.01749
CLD	15	0.00390
Total cost/pc		0.09065 THB
		$0.00254 USD

Note: $1 USD = 35.7 THB

Table 8. Physical properties of the best formula comparing with TIS 614-1986 standard.

	Bending Strength (Mpa)	%Water absorption
TIS 614-1986 Standard	20	6
Formula No.4_1	2.32	3.2

As shown in Table 8, bending strength of the best formula is lower than that of to the TIS 614-1986 standard. However, its water absorption is approximately close that standard. Enhancement the bending strength should be further investigated. The first aspect is to study about extending the curing period of specimens from 7 days to 28 days [6]. In addition, the second aspect is to utilize the other alternative materials such as waste ash from boiler of palm oil mill industry. Because this material composition has the silica content which can be replaced in laterite soil and fluvial sand [13]. Furthermore, increasing the composition of Portland cement in the formula should be considered [10].

Although the appropriate formula could not achieve the TIS standard's bending strength. This product should be used as eco-friendly product utilizing the waste materials. In addition, it could be substitute the building material that needs not to be fired. In the future, if bending strength of products can meet TIS 614-1986, it can response the customers who want to use the eco-friendly products.

References

1. Department of Primary Industries and Mines, Ministry of Industry, Access on : June 2, 2015 http://www1.dpim.go.th/csh/cr.php
2. Sila Chairoen Company, Access on : June 5, 2015 https://www.facebook.com/silachaicharoen?fref=nf
3. M.Galetakis, S. Raka, Utilization of limestone dust for artificial stone production: an experimental approach, Minerals Engineering. **17**, 355–7 (2004)
4. A.A. Mageeda, G.S. AbdelHaffez,Utilization of Limestone Dust in Brick Making, Journal of Engineering Sciences, Assiut University. **40**, 3 913-922 (2012)
5. I. S. Fathi, Effect of Using Crushed Limestone in Concrete Mixes as Fine Aggregate on Compressive Strength and Workability, International Conference on Civil, Biological and Environmental Engineering (CBEE-2014). Istanbul (Turkey) (2014)
6. P.Turgut, M. A. Halil, Limestone dust and wood sawdust as brick material, Building and Environment. **42**, 3399–403 (2007)
7. P. Turgut, Cement composites with limestone dust and different grades of wood sawdust, Building and Environment. **42**, 3399–3403 (2007)

8. P. Turgut,Limestone dust and glass powder wastes as new brick material, Materials and Structures. **41**, 805–813 (2008)
9. P. Turgut, Research into Artificial Limestone Composites with Cotton Waste, Jordan Journal of Civil Engineering. **7**, 1 126-131 (2013)
10. U. Wangrakdiskul, P. Khonkaew, T. Wongchareonsin, Use of the Spent Bleaching Earth from Palm Oil Industry in Non Fired Wall Tiles, The 3rd International Symposium on the Fusion Technologies 2014 (ISFT2014), Jeonju, S.KOREA, 30 July-3 August. (2014)
11. U. Wangrakdiskul, T. Kleepung, P. Srihakham, The Possibility Study of Using Laterite Clay, Soil and Cement Interlocking for Non Fired Wall Tiles, Proceedings of Industrial Engineering Network Conference 2008. 20th-22ndOctober. pp10-15.Songkhla. Thailand. (2008)
12. M.J. Anderson, S.L. Kraber, Keys to successful designed experiments. Tech. Rep., Stat-Ease, Inc., Minneapolis, MN (2002). [Online]. Available: http://www.statease.com/pubs/doe-keys.pdf
13. V. Sata, C. Jaturapitakkul, K. Kiattikomol, Utilization of Palm Oil Fuel Ash in High-Strength Concrete, Journal of Mater

Nonlinear Progressive Collapse Analysis Including Distributed Plasticity

Osama Ahmed Mohamed, Zubair Syed Imam and Rania Khattab

College of Engineering, Abu Dhabi University, United Arab Emirates

Abstract. This paper demonstrates the effect of incorporating distributed plasticity in nonlinear analytical models used to assess the potential for progressive collapse of steel framed regular building structures. Emphasis on this paper is on the deformation response under the notionally removed column, in a typical Alternate Path (AP) method. The AP method employed in this paper is based on the provisions of the *Unified Facilities Criteria – Design of Buildings to Resist Progressive Collapse*, developed and updated by the U.S. Department of Defense [1]. The AP method is often used for to assess the potential for progressive collapse of building structures that fall under Occupancy Category III or IV. A case study steel building is used to examine the effect of incorporating distributed plasticity, where moment frames were used on perimeter as well as the interior of the three dimensional structural system. It is concluded that the use of moment resisting frames within the structural system will enhance resistance to progressive collapse through ductile deformation response and that it is conservative to ignore the effects of distributed plasticity in determining peak displacement response under the notionally removed column.

1 Introduction

Progressive collapse of building structures is not covered extensively in international building codes and design standards. The U.S. Department of Defense (DoD) publishes UFC 4-023-03, a guide for progressive collapse resistant design of buildings structures [1].

A building structure is assessed for progressive collapse potential, after it is designed in accordance with the applicable building code. The structural system is assessed in terms of its capacity to transfer loads in ductile manner, after the loss of primary load-carrying members. Framed structures are assessed by quantifying the consequences of the notional removed of critical corner, interior, and exterior columns. Loss of corner columns is particularly critical because its leads to long unsupported spans [2].

The current UFC guide [1] embraces the design philosophy that that progressive collapse resistant steel structures are ductile systems, similar in some ways to earthquake resistant structural systems, particularly when certain types of moment resisting connections are used [3]. Several researchers suggested that lessons learned from decades of studying structural performance during earthquakes that aim at producing ductile structural systems could guide engineers in producing progressive collapse resistant systems [4]. UFC defines the required strength for each structural component as force-controlled or deformation-controlled, based on the force-deformation relationship for the required strength.

This paper outlines the application of the AP method to assess the capacity of steel structural systems to transmit gravity load, following the loss of a primary load carrying member in a framed steel structural system using nonlinear static analysis procedures. Dynamic effects associated with the sudden removal of column are incorporated using load increase factors. Linear analysis procedures are the dominant methods in professional design offices for a large class of structures, with the exception of geometric nonlinearities in steel structures that must be rigorously investigated. The application of the AP method using Linear Static Analysis (LSA) procedures is permitted for structures that do not contain significant irregularities or if the demand-to-capacity ratio (DCR) is less than 2.0. The computationally more expensive Nonlinear Static Analysis (NSA) is permitted without limitations on DCR or structural irregularities. The use of LSA and NSA with the AP method is permitted by UFC [1] and is justified in two ways: 1) The computational cost of LSA and NSA is lower than dynamic analysis, 2) progressive collapse is hazard independent and involves significant uncertainties that justify using approximate methods such as LSA and NSA. This paper is based upon Mohamed et al. [5], which examines the use of LSA procedures for AP investigations. The current paper employs NSA to assess the potential for progressive collapse of steel framed structures and addressed the effect of ignoring distributed plasticity on deformation response.

2 Nonlinear static analysis procedure and the alternate path method

The alternate path method is one of the most commonly used procedures to assess the ability of the structure to transmit gravity loads safely to competent foundation. The requirement to use the AP method depends on the building Occupancy Category (OC). AP method is typically required for OC III and OC IV, and in some cases for OC II. OCs are defined in UFC 3-301-01 Structural Engineering [6]. Occupancy categories defined in UFC 3-301-01are similar to the Risk Categories currently used in ASCE 7 – 10 [7].

2.1 Gravity load for assessment of progressive collapse potential

Progressive collapse is a gravity driven extreme event. In order to assess the vulnerability of a building structure to progressive collapse, alternate path analysis is conducted after the notional removal of selected columns [1]. Prescribed gravity loads are applied and force-controlled and deformation-controlled actions are determined. These actions are compared to certain acceptance criteria, which may be incorporated in the analysis software.

Deformation-controlled and force-controlled actions on the structural system obtained using NSA procedure are calculated by applying amplified gravity loads, given by Eq. 1 to the floor panels adjacent to the notionally removed column at all floor levels [1,5].

$$G_N = \Omega_N [1.2D + 0.5L] \qquad (1)$$

where,

G_N : Increased gravity loads for deformation controlled actions calculated using NSA method.

D: Service dead loads including façade (kN/m2)

L: Service live load including applicable reductions, per ASCE 7-10 (kN/m^2)

Ω_N : Dynamic increase factor for calculating deformation-controlled and force-controlled actions using NSA method.

Eq. 2 represents the gravity loads on floor panels that are not adjacent to notionally removed columns, and used to determine either deformation-controlled or force-controlled actions, as appropriate. The gravity loads in Eq. 1 and Eq. 2 are applied simultaneously at all floor levels.

$$G = 1.2D + 0.5L \qquad (2)$$

The gravity load for LSA procedures is similar to NSA, except that a Load Increase Factor (LIF) is used instead of the dynamic increase factor indicated in Eq. 1. Progressive collapse assessment using the alternate path method and LSA procedures as well as the process for determination of LIF are discussed elsewhere [1, 8]. The discussion in this paper is limited to deformation-controlled actions.

3 Dynamic increase factor

In this study, the AP method was applied using nonlinear static analyses, where the dynamic effects of load removal are approximated through the use of dynamic increase factors (DIF) as shown in Eq. 1.

For steel framed structures, the dynamic increase factor (DIF), Ω_N to be used in Eq. 1 is defined as $\Omega_N = 1.08 + 0.76/(\theta_{pra}/\theta_y + 0.83)$, in which θ_{pra} is the plastic rotation angle, and θ_y is the yield rotation [9].

4 Acceptance criteria

For deformation-controlled actions calculated using NSA, the acceptance is not based on component strength, but rather on deformation limit controlled by acceptance criteria [1]. The post-yield behaviour is simulated in this paper by inserting concentrated plastic hinges on steel frame objects. The hinge model used in this study to conduct NSA is shown in Fig. 1.

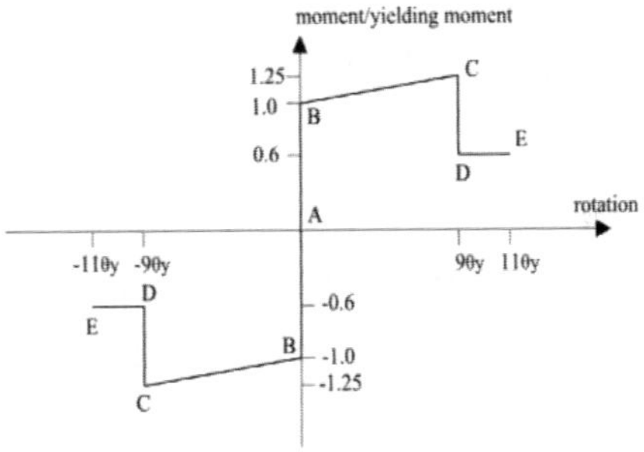

Figure 1. Nonlinear hinge model for nonlinear static analysis.

The plastic hinge parameters and acceptance criteria for selected shapes are shown in the table below [9]. The w-shapes shown below represent some of the design sections for the case study discussed in a subsequent section of this paper.

Progressive collapse is an extreme event; therefore, the acceptance criteria used for NSA and applied to deformations is based on collapse prevention, as recommended by [1].

NSA conducted in this paper for progressive collapse assessment, assumes plastic hinges described in Fig. 1 is assigned to frame members. The plastic hinge length is as per FEMA 356 [9].

5 Case study

In order to evaluate the effect considering or ignoring distributed plasticity when conducting nonlinear static analysis during AP investigation, the case study developed by Mohamed and Tarmoom [5] is adopted. The building structure described by Fig. 2 and Fig. 3 was analysed and designed. The floor system is a reinforced concrete slab on a steel deck. All secondary beams were designed as composite beams while moment resisting frames were designed without composite action. Ordinary moment resisting frames exist along the perimeter of the building and along grid lines A to H.

The case study conservatively ignores the contribution of the deck to progressive collapse resistance. Consideration of composite action in assessment of progress collapse was discussed in the literature [10]. The member shown in Fig. 2 and Fig. 3 were sized for reducible live load including light weight partitions of 3.6 kN/m^2. The weight of the concrete-filled steel deck was 2.79 kN/m^2. Superimposed dead load was assumed to be 4 kN/m^2, along with cladding load of 3.2 kN/m. The relatively high gravity load ensures progressive collapse is indeed gravity driven. Design wind speed was 161 km/hr.

The steel framed structure was designated as OC III and was analysed and designed according to AISC LRFD specifications [11], using the Direct Analysis Method. The longest span of 10 meter was chosen so as to reflect the effect of long spans in exacerbating progressive collapse.

Beam	Plastic Rotation Angle		Residual Strength Ratio	Acceptance Criteria		
				Plastic Rotation Angle, Radians		
	a	b	c	Immediate Occupancy	Life Safety	Collapse Prevention
W610x113 (W24x76)	9θy	11θy	0.6	1θy	6θy	8θy
W610x372 (W24x250)	9θy	11θy	0.6	1θy	6θy	8θy
W530x138 (W21x93)	9θy	11θy	0.6	1θy	6θy	8θy

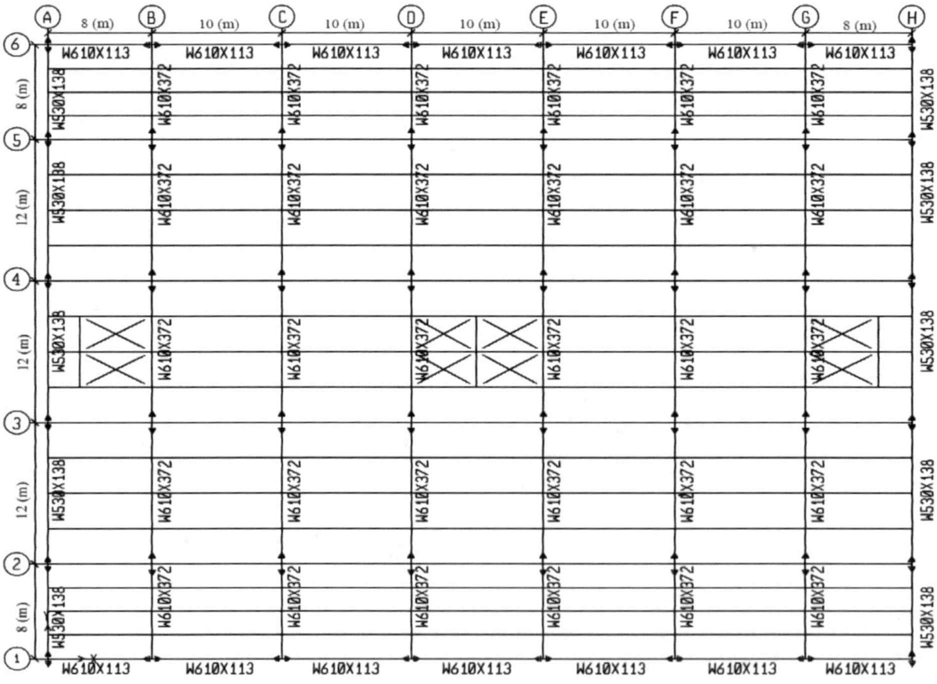

Figure 2. Typical plan view of case-study 15-story steel building.

Design was conducted using the computer program ETABS, developed by Computers and Structures Inc., USA. Three-dimensional models were used for analysis/design as well as progressive collapse investigations to avoid overly conservative structural response. It should be noted that ignoring three-dimensional effects is not conservative when beams frame into exterior frames as they may induce considerable torsional effects, especially in corner columns as demonstrated by Mohamed [2]. Therefore, all interior moment frames along grid lines A to H were assumed to frame into perimeter frames along grid lines 1 and 6 using shear connections. This will theoretically eliminate torsional moments on perimeter frames.

6 Distributed plasticity

Mohamed and Tarmoum [5] applied the AP method to assess the potential for progressive collapse of the case study described in this paper. When column D2 was notionally removed, the maximum vertical deformation under this columns was 1342.60 mm, when using nonlinear static analysis using the load combinations in Eq. 1 and Eq. 2. All of the designed w-shaped sections in Fig.2 and Fig. 3 meet the requirements of section I2.2 of the AISC specifications [10] for prevention of local buckling. The requirements of section I2.3 [10] to control lateral torsional buckling are satisfied through the concrete-filled composite deck on top of the w-shaped beams. In order to understand the effect of distributed plasticity on the deformation response under notionally removed column, additional concentrated hinges are applied along vertical and horizontal frame elements. The results are summarized in the following subsections.

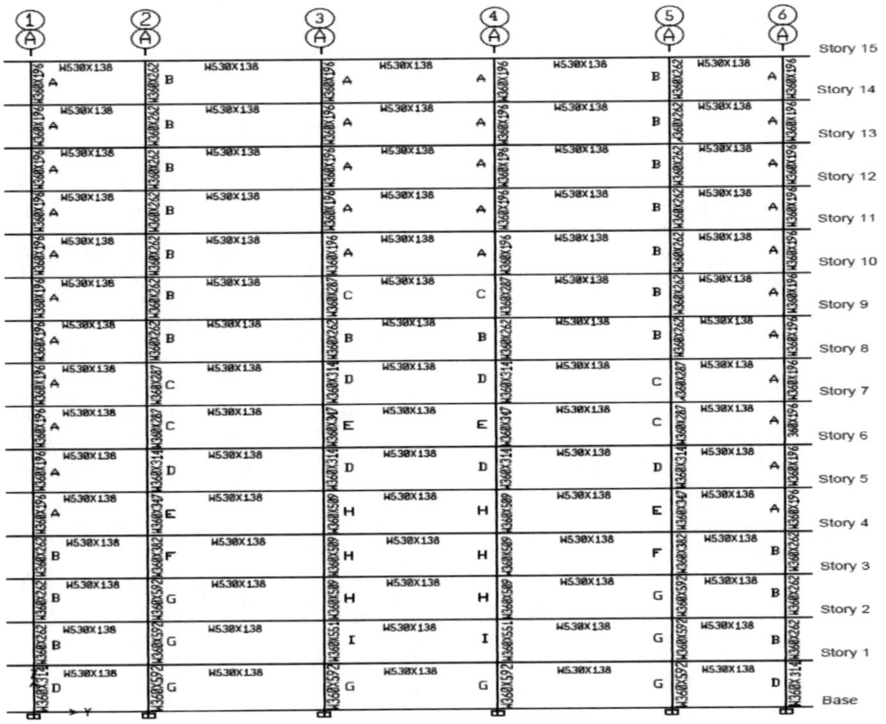

Figure 3. Elevation along grid line A, beams sections are all W530X138, column sections are; A: W360X196, B: W360X262, C: W360X287, D: W360X314, E: W360X347, F: W360X382, G: W360X592, H: W360X509, I: W360X551.

Analysis Results for Notional Removal of Interior Column D2 – Three plastic hinges assigned at the begining and at the end of each horizontal and vertical element

Mindful of the computational cost, only three plastic hinges are used at the end of each frame element to capture the effect of distributed plasticity on deformation response. When column D2 is notionally removed, the unsupported span after removal of column D2 reaches 20 meters. Analysis of the structural system with column D2 removed produced the deformed shape shown in Fig. 4 and the maximum vertical deformation response reaches 1934.6 mm. It is noted that yielding was limited to part of the floor adjacent to the notionally removed column D2 and didn't spread to other areas of floor.

Load Increase Factor for Interior Column D2 - Five plastic hinges assigned at the begining and at the end of each horizontal and vertical element

Mindful of the computational cost, number of plastic hinges is increased to three that are applied at the end of each frame element to capture the effect of distributed plasticity on deformation response. Analysis of the structural system with column D2 removed produced the deformed shape shown in Figure 5 and the maximum vertical deformation response reaches 955 mm. Analysis of the structural system with column D2 removed produced the deformed shape shown in Fig. 5. Similar to the case of 3-plastic hinges, yielding remained within the beams surrounding the notionally removed column, and damage is essential restricted to the floor area surrounding the notionally removed column. Fig. 5 shows additional plastic hinges yielding on all beams D1-D2 at each story level, compared to the case of fewer modelled plastic hinges shown in Fig. 4. Furthermore, additional hinges formed on beams D2-D3 at the upper four stories when 5-hinges are modelled as shown in Fig. 5, compared to Fig. 4.

Figure 6 shows the members assessed for stress level based on DCR, for the model where five plastic hinges were used on each frame element. Only the two members connected to the notionally removed column failed in each floor along grid line D, namely, beams D1-D2 and D2-D3. This is because beam D1-D2 and D2-D3 became one 20-meter long beam.

7 Summary

The use of moment frames along the perimeter and interior of steel framed structure enhances the resistance to progressive collapse through limiting the response to the floor areas near the notionally removed columns. Moment resisting connections are more expensive to build compared to simple connections that are typically used with gravity systems. When notionally removed internal column is within properly design moment frame, ductile behavior may be expected during progressive collapse AP analysis, even with relatively long beams spans, similar to the ones examined in this study.

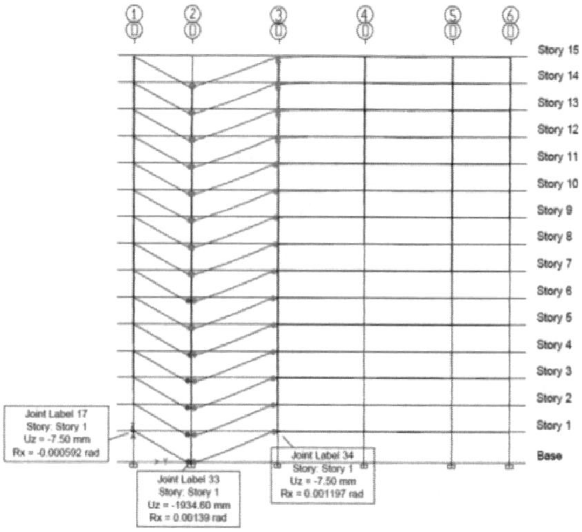

Figure 4. Deformed shape with notionally removed column D2 and parallel interior moment frames-NSP.

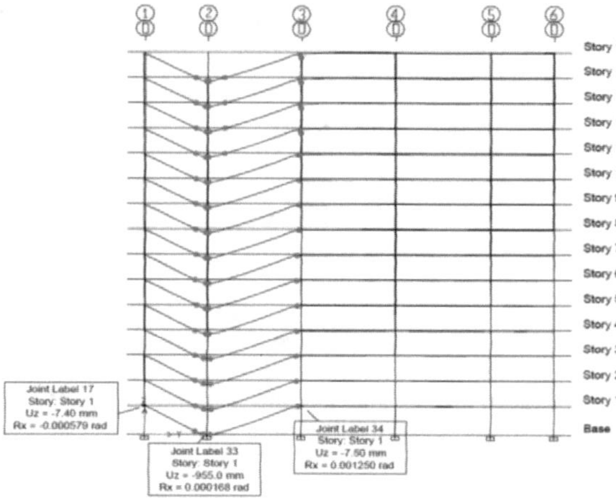

Figure 5. Deformed shape with and plastic hinges when D2 is notionally removed - each frame element modelled with five distributed hinges along.

Considering distributed plasticity using an acceptable method improves accuracy of responses obtained from AP progressive collapse analysis. It is conservative to ignore distributed plasticity in computing deformation responses. A simple approach to account for distributed plasticity that involves modelling several concentrated hinges at suitable locations may offer sufficient accuracy to predict the desired deformation-controlled behaviour. Computationally expensive models for distributed plasticity are not always necessary, particularly when considering the high level of uncertainty involved in progressive collapse assessment methods.

The findings of this study are applicable to steel structures where the lateral force resisting system consist of moment frames only, without bracing system.

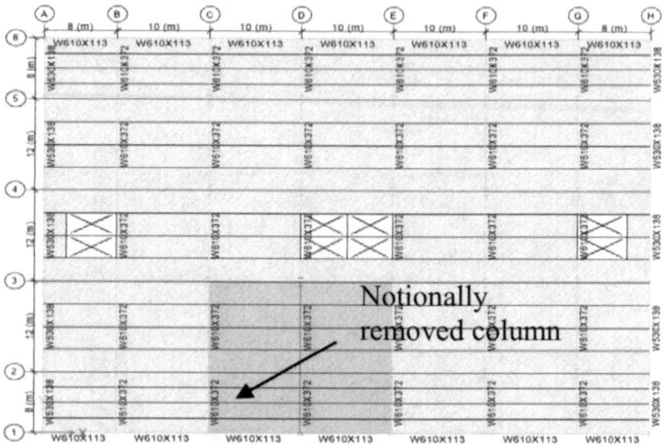

Figure 6. Beams D1-D2 and D2-D3 exceeded DCR of 1.0 due to notional column removal when each frame element is modelled with five distributed plastic hinges.

Acknowledgement

The authors gratefully acknowledge the financial support of the Center on Sustainable Built Environment and the Office of Research and Sponsored Programs at Abu Dhabi University.

References

1. U.S. Department of Defense, Design of buildings to resist progressive collapse, UFC 4-023-3, Unified Facilities Criteria, U.S. Department of Defense, 2013.
2. O. A. Mohamed, Assessment of Progressive Collapse Potential in Corner Panels of Reinforced Concrete Buildings, Journal of Engineering Structures, **31**, 3 749-757 (2009)
3. T. Kim and J. Kim, Collapse Analysis of Steel Moment Frames with Various Seismic Connections, Journal of Constructional Steel Research, **65**, 1316-1322 (2009)
4. K. Khandelwal, S. El-Tawil, F. Sadek, Progressive Collapse Analysis of Seismically Designed Steel Braced Frames, Journal of Constructional Steel Research, **65**, 3 699–708 (2009)
5. O.A. Mohamed, M.G. Tarmoom, Alternate Path Method for Progressive Collapse Resistant Design of Steel Framed Structures, in J. Kruis, Y. Tsompanakis, B.H.V. Topping, (Editors), Proceedings of the Fifteenth International Conference on Civil, Structural and Environmental Engineering Computing, Civil-Comp Press, Stirlingshire, UK, (2015)
6. U.S. Department of Defense, Structural Engineering - UFC 3-301-01, Unified Facilities Criteria, U.S. Department of Defense, (2014).

7. American Society of Civil Engineers, Minimum Design Loads for Buildings and other Structures ASCE/SEI 7-10, ASCE, (2010)
8. O. A. Mohamed, Calculation of load increase factors for assessment of progressive collapse potential in framed steel structures, Case Studies in Structural Engineering, 3 11-18 (2015)
9. American Society for Civil Engineers, Seismic Rehabilitation of Existing Buildings ASCE/SEI 41 – 6, ASCE, (2007).
10. Y. Alashker, S. El-Tawil, and F. Sadek, Progressive Collapse Resistance of Steel-Concrete Composite Floors, Journal of Structural Engineering – ASCE. **136**, 10 (2010).
11. American Institute of Steel Construction, Steel Construction Manual – AISC 325 – 11 - 14th Edition, 3rd printing, AISC (2011)

Nonlinear Progressive Failure Analysis of Surrounding Rock System Based on Similarity Theory

Y. Zhao[1,2], and L. He[1]

[1]College of Civil Engineering, Chongqing University, China
[2]Key Laboratory of New Technology for Construction of Cities in Mountain Area(Chongqing University), Ministry of Education, China

Abstract. Nonlinear progressive failure study of surrounding rock is important for the stability analysis of underground engineering projects. Taking a deep-buried tunnel in Chongqing as an example, a three dimensional(3-D) physical model was established based on similarity theory. To satisfy similarity requirement of physical–mechanical properties, such as elastic modulus, compressive strength and Poisson ratio, physical model materials were developed. Using full inner-spy photograph technology, the deformation and failure process of rock were studied under the situation of independent and combined action of anchor, shotcrete and reinforcing mesh. Based on experimental results, the interaction mechanism between rock and support structure under high stress was investigated.

1 Introduction

Rock system is a typical multi-components material, which consists of rock block, fracture and other components. Under the deep buried conditions, rock deformation may not belong to elasticity range, the interaction mechanisms of physical nonlinear and geometric nonlinear will make the evolution of rock system extremely complicated [1-3]. Similarity theory has been proved to be very useful to solve the problem [4-5]. As a significant method of science research, physical model test is the first introduced model which similar to the prototype in laboratory based on the similarity principles, by observing mechanical parameters and distribution of model with instrument, then deduce possible mechanics phenomenon and pressure distribution of the prototype with results of model, finally solve practical problems of rock engineering. This research method is simple, intuitive, economic and fast, as well as its period is short.

Using model test, displacement back analysis and boundary element calculation, Chi Yong et al. studied the stability of full seam gateway supported with anchor, and his research reflected good combination of model test with numerical calculation [6]. Chen Anming et al. developed a creep equation for soft rock materiel to study the model test of soft rock reinforced anchor cable, and discussed the variation of tension force of anchor cable with time [7]. Zhu Fangcai et al. studied rock burst character of similar material failure under different stress paths for underground tunnel [8]. Li Zhongkui et al. did a lot of meaningful researches on 3-D model simulation technology [9]. However, some strength indexes (like cohesive force, internal friction angle, tensile strength), the physical and mechanical properties of post-peak rock (like residual cohesive force, residual internal friction angle) are considered rarely, which makes it difficult to identify the failure mechanisms of rock mass [5].

This paper studied the stability of surrounding rock for deep-buried long tunnel by physical model test on true triaxial apparatus designed by Institute of Rock and Soil Mechanics, Chinese Academy of Sciences. Based on available materials, new similarity materials were compounded by ameliorating present prescription. The indexes of physical and mechanical properties of materials compounded in this study, such as compression strength, tensile strength, Poisson's ratio, internal friction angle, meet the requirement of similarity relation, stimulate high geostress field of deep-buried tunnel, perform the 3-D physical model tests of three-centered arch cavity under the situation of independent and combined action of anchor, shotcrete and reinforcing mesh. Based on the experiment results, the paper studied the interaction mechanism between the evolution of rock deformation and support structure under high stress.

2 Similarity theory

In order to make the model and prototype have similar physical phenomena, materials, shape and load of model must follow the similarity principles of the prototype. For geomechanics model test, it must make the equilibrium equation, compatible equation, geometric equation, physical equation and boundary condition of model fully agree with the prototype, it also make the strain of model is the same as the prototype, and strength criterion and stress-strain relationship of model similar to the prototype. These similarity coefficients which defined as the ratios of prototype parameters to model parameters must be constants. In this work, subscripts p and m denote prototype parameters and physical models parameters, respectively, and then these similarity constants can be defined as:

$$C_l = \frac{l_p}{l_m}, C_v = \frac{v_p}{v_m}, C_\gamma = \frac{\gamma_p}{\gamma_m} \tag{1a}$$

$$C_\sigma = \frac{\sigma_p}{\sigma_m} \ C_E = \frac{E_p}{E_m} \ C_\varepsilon = \frac{\varepsilon_p}{\varepsilon_m} \tag{1b}$$

$$C_C = \frac{C_p}{C_m}, \ C_\varphi = \frac{\varphi_p}{\varphi_m} \tag{1c}$$

where C_l, C_v, C_γ, C_σ, C_E, C_ε, C_C and C_φ indicate similarity constants for geometry, Poisson's ratio, density, stress, elastic modulus, rain, cohesion force and internal friction angle similarity constant ,respectively.

And following similarity equations are described as:

$$C_E = \frac{C_\sigma}{C_\varepsilon} \tag{2}$$

$$\frac{C_\sigma}{C_l C_\gamma} = 1 \tag{3}$$

$$\frac{C_c}{C_l C_\gamma} = 1, \frac{C_\varphi}{C_l C_\gamma} = 1 \tag{4}$$

$$C_\varepsilon = 1, \ C_\gamma = 1 \tag{5}$$

In practical application, it is difficult to get complete similarity model which all similarity principles are satisfied, so the model just need to meet the main similarity principles.

All aspects should be considered for the test accuracy, such as the test load and economic indexes of the model, suitable model materials which are difficulty to search, the test technology and simulation range of existing equipment. Therefore, it should analysis multiple schemes of model scale to ensure the accuracy, then the parameters finally selected are shown in Table 1.

Table 1. Similar scale of model test.

Similarity constant	Value
C_l	50
C_σ	40
C_E	40
C_c	1
C_ϕ	1
C_v	1

3 Simulation of prototype model

The physical model size is 80cm×80cm×80cm(as shown in Fig. 1). The model is poured by four stratums, and thickness of each stratum is 20 cm. Anchors are established in the second, third and fourth stratums. The most length of long anchor in the second and third stratums is 82 mm, while short anchor is 55 mm. In the fourth stratum, the length of long anchor is 110 mm, while short anchor is 91 mm. Lining is established in the third and fourth stratums (as shown in Fig. 2). Lining is established in the third and fourth stratums, the thickness is 5 mm (as shown in Fig. 3).

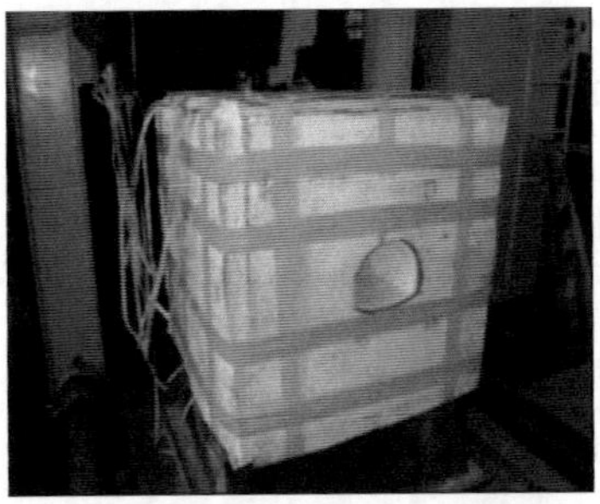

Figure 1. Three-dimension model.

Figure 2. Anchor arrangement.

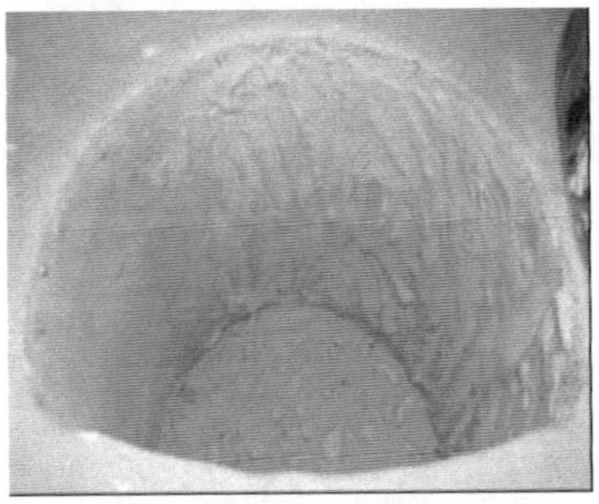

Figure 3. Initial lining.

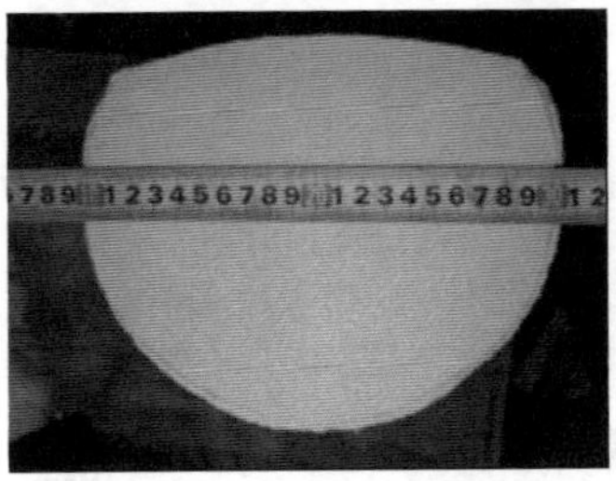

Figure 4. Thickness of initial lining.

DOI: 10.1051/matecconf/20166407004

4 Model test and results analysis

4.1 Steps of loading

The ratio of horizontal load to vertical load in this test is 0.8. The horizontal load (X direction) and the vertical load (Z direction) start with 200 KN and 250KN, respectively, and then are elevated by 40KN and 50KN step by step, respectively. When elevated to 400 KN and 500 KN, the load for horizontal load per step is elevated by 20 KN and vertical load by 25 KN. When the horizontal and vertical load is increased to 1120 KN and 1400 KN, respectively, stop loading.

4.2 Results analysis

(1) The first stratum: cavity without any support

Middle vault collapse is the main failure form for cavity, and the thickness of collapse pieces is about 20 mm. The bottom of arch is generally in good conditions. And the horizontal displacement of sidewalls is dominant for the surrounding rock; the maximum value is about 7~8 mm appearing in the area between arch waist and arch feet (as shown in Fig. 5).

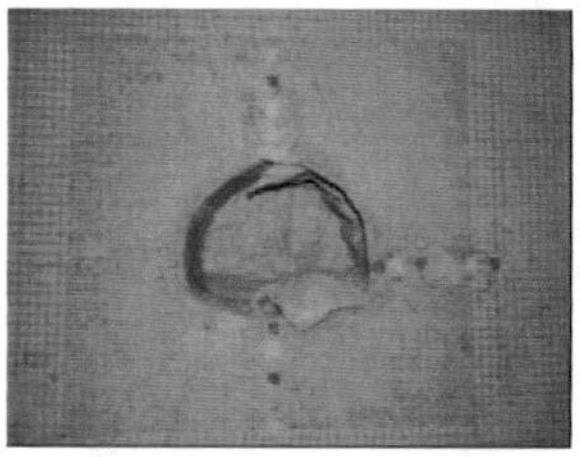

Figure 5. Failure of the first stratum.

The height of disturbed zone is about 0.5 D (D indicates the radius of the cavity). Vertical downward displacement occurs above arch waist (including surrounding rock of the vault), and the height of affected zone in vault is about 0.6 D. The maximum displacement is at vault and the value is about 3 mm. Vertical upward displacement occurs below the bottom of arch, but the affected zone is roughly 0.1~0.2D smaller. Two big cracks and a fractured zone occurred in surrounding rock, one of the two big cracks develops from the collapse area of vault to the model boundary, while the other crack develops from the collapse area of arch feet to the model boundary is about 45°. Fractured zone appears in the area from arch waist to arch feet, the deepest thickness of these zones is about 25 mm.

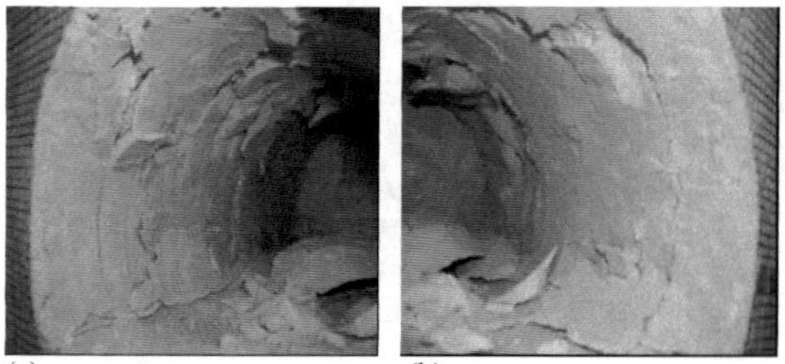

(a) (b)

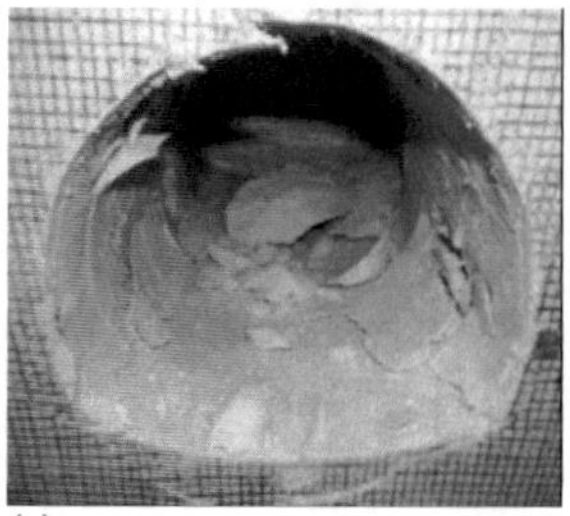

(c)

Figure 6. Failure of the second stratu.

(2) The second stratum: anchor only supportsarch waist applied above

Vertical rhegma of sidewalls is the main failure form. Loading causes spalling, the thickness of chunks of rock seperating is about 8 mm in the position of arch foot. Vault and bottom of arch are nearly intact (as shown in Fig. 6).

The horizontal displacement is the main deformation of surrounding rock, and the maximum value which appears at arch feet is about 5 mm. The deformation is focused on some regions nearby arch feet, and the depth influenced is about 0.3 D. The deformation above arch feet is not obvious. The maximum vertical upward displacement occurring in arch feet is maximum vertical displacement and the value is about 5 mm. The vault's vertical downward displacement is relatively small, and the value is about 1~2 mm. In addition, a vertical crack develops from vault to model boundary.

Figure 7. Failure of the third stratum.

Figure 8. Failure of the forth stratum.

(3) The third stratum: On the basis of the second stratum, a lining is established above arch feet. A fine iron wire net is placed in lining.

The horizontal displacement is dominant for the surrounding rock, and the maximum value appears in arch waist, is about 2~3 mm. The vertical displacement is small. Some small pieces of bursting cracks occur in bottom of arch (as shown in Fig.7). Spray layer is nearly intact, except a leap (the fine iron wire net completely exposes) parallel cavity axis appears in the middle part of a side between vault and arch waist (as shown in Fig.7). In addition, two approximately paralle vertical cracks occur in one side of vault.

(4) The fourth stratum: anchor support arch waist applied above, and the length of anchors is longer than the third. The anchors of 55 mm in the third stratum are lengthened to 91 mm, while anchors of 82 mm are lengthened to 110 mm. And the spray layer is the same as the third stratum

There is no failure in the surrounding rock. And the spray layer does not break off from surrounding rock (as shown in Fig.8).

5 Discussion and conclusions

The deformation, failure depth and affected region in the first stratum are the most outstanding, but the bottom of arch remains nearly intact. In the second stratum, the failure region is less than the first stratum, and only a small slump occurs in the surface area of the vault, the tangential stress of side walls decreases significantly. It shows that mechanical anchoring effect increases with the decreasing of disturbed zone. In the third stratum, the main failure of cave walls is the crack at the bottom of arch and the leap of spray layer; the other depth failure is less. The fourth stratum has no failure phenomenon, it is more stable than first three stratums.

Comparing among these four stratums, it can conclude that failure deformation of cavity without any support is the largest. Many rock units are in the post-failure regions according to the peak stress. The residual strength is determined by friction between strips. It develops a new rock of mechanics boundary, which makes rock's property close to loose massive media, and determines rock pressure and residual strength. The existence of the residual strength has great influence on stability of deep-buried tunnel under high rock pressure.

After applying anchor to support arch waist, the stability of cave walls is good. There is almost no axial crack and fall-block, and the cracks mainly occur in weak section between anchors. Large cracks connecting the two arch feet occur in the bottom of arch with no any reinforce measures, which belongs to integral failure. When support is provided by rock bolts and shotcrete, failure occurs in the spray layer and the bottom of arch, but the surrounding rock is nearly intact, further increase the length of anchors, the stability of the surrounding rock is in a good condition.

Acknowledgments

This work was supported by the New Century Excellent Talent Foundation from MOE of China(Grant No. NCET-09-0844), the Fundamental Research Funds for the Central Universities (Grant No.CDJZR11 20 00 14), and National Natural Science Foundation of China (Grant No.50804060).

References

1. Shi X.Z., Zhou J., Li X.B., Utilization of a nonlinear support vector machine to predict blasting vibration characteristic parameters in opencast mine, Przegląd Elektrotechniczny, 88 127-132 (2012)
2. Ren L., Zhao J.Z., Hu Y.Q., Wang L., Effect of natural fractures on hydraulic fracture initiation in cased perforated boreholes, Przegląd Elektrotechniczny, 88 108-112 (2012)
3. Zhao Y., Peng H.Y., Analysis of nonlinear dynamic character in the surrounding rock system for deep buried underground engineering, Journal of Coal Science and Engineering (China), **16**, 4 362-366 (2010)
4. Liu J., FengX.T., DingX.L., ZhangJ., Yue D.M., Stability assessment of the Three-Gorges Dam foundation, China,using physical and numerical modeling-Part I: physical model tests, International Journal of Rock Mechanics & Mining Sciences, **40**, 5 609-631 (2003)
5. Zhao Y., Peng H.H., Experimental study on similarity materials for soft rock of deep-buried tunnels, Journal of Coal Science & Engineering (China), **17**, 2 124-127 (2011)
6. Chi Y., Zhou J., Kang J.H., Simulation analysis of stability of coal roadway bolting, Mining and Metallurgical Engineering, **22**, 1 39-41 (2002)
7. Chen A.M., Gu J.C., Shen J., Ming Z.Q., Model testing research on the variation of tension force of anchor cable with time in reinforcement of soft rocks, Chinese Journal of Rock Mechanics and Engineering, **21**, 2 251-256 (2002)
8. Zhu F.C., Study on failure models of simulated material under different stress paths, Journal of Central South University of Technology, **33**, S 8-11 (2002)
9. Li Z.K., Lu D.R., Nakayama H., Hosomi H., Sun J.S., Development and application of new technology for 3d geo-mechanics model test of large underground houses, Chinese Journal of Rock Mechanics and Engineering, **22**, 9 1430-1436 (2003)

Differential Diffusion Effects in Numerical Simulations on Smoke Toxicity Evaluation

Longwei Pan, Yong Jiang, Lei Wang, Lulu Yin and Li Xu

State Key Laboratory of Fire Science, University of Science and Technology of China, Hefei, China

Abstract. With the increasing frequency of fire caused by construction materials, smoke toxicity evaluation plays a key role in related fields. Numerical simulation has become a popular method to predict the toxicity of smoke. A computational study of differential diffusion effects on smoke toxicity evaluation is proposed in this study. Further, an effective Lewis number model derived from the Reynolds-averaged form of the Navier-stokes (RANS) transport equations is proposed in turbulent flames. The accuracy of the study is illustrated for a polyurethane foam fire in a 1/5 scale vertical shaft. The temperature and the concentrations of smoke composition are mainly discussed. From the comparison of the calculations with the direct numerical simulations (DNS) data it is observed that the temperature and mass fractions of species agree well with the DNS data when differential diffusion effects are taken into account. On the other hand, these numerical results are overestimated if differential diffusion effects are neglected. The FED values indicate that differential diffusion has a strong influence on smoke toxicity evaluation when using N-Gas model.

1 Introduction

The number of high-rise buildings is increasing rapidly with social and economic development and urbanization in recent years. Meanwhile, the demand for construction materials has been greatly increased because of the large number of architectures. Unfortunately, most of the construction materials will bring serious fire risks. According to previous researches, the frequency of fire caused by building materials is increasing year after year [1, 2]. Toxic smoke was identified to be the most hazardous factor to human beings in the fire caused by construction materials [3, 4, 5]. Apparently, smoke toxicity evaluation plays a key role in related fields.

At present, there are several methods [6] for testing the toxicity of flue gas: (1) small scale test method. This method does not have a unified standard. In China, the smoke production principle follows the German standard [7]; (2) animal infected test method. Animal infected test was mainly used in the past. Usually there are several measurements, LC50 (Lethal Concentration), IC50 (Incapacitation Concentration), EC50 (Effete Concentration) and so on, and LC50 is the most commonly used [8, 9]; (3) component analysis method. Component analysis is mainly based on different characteristics of the fire smoke components. Most of the methods can only analysis the intermittent sampling process, not the whole combustion process [10]. The final toxicity values can be obtained by comprehensive evaluation after the detection of composition and concentration of gas by the above method. The evaluation models of flue gas toxicity mainly have N-Gas, FED, FEC and TGAS model [11]. N-Gas model is considered to be one of the most precise and scientific one [12]. There is an increasing tendency to use numerical simulation method instead of experimental method to test the toxicity of gas to prevent the use of animal testing for all except some very limited applications [13]. Therefore, it is very necessary to predict the composition and concentration of gas accurately.

Nowadays, CFD (Computational Fluid Dynamics) simulation has become a very popular and convenient method to research building fires, especially to predict gas concentrations. As one of the CFD software, FDS (Fire Dynamics Simulator) is widely used to solve fire cases. In FDS, large eddy simulation is a more practical method to simulate building fires. But large eddy simulation adopts the assumptions of equal thermal and mass diffusivities, leading to the unity Lewis number for all combustion products of materials [14]. Unfortunately, species found in industrial applications rarely have unity Lewis numbers [15]. Actually, species and heat locally redistribute (this phenomenon is referred to as differential diffusion [16]) when Lewis number is not the unity [17]. Moreover, the species concentrations and temperatures are quite different when differential diffusion effects are taken into account [18].

In this paper, FDS is chosen to predict the gas composition and concentration in a fire and a study on effects of differential diffusion on smoke toxicity is implemented by modifying FDS code. The following content are mainly divided into four parts. First, governing equations and computational methodology are introduced. The open source program is compiled after modifying the thermal and mass diffusivities formulas. In addition an effective Lewis number model derived from the Reynolds-averaged form of the Navier-stokes (RANS) transport equations is proposed. Second, a 1/5 scale vertical shaft model [19] is used as the physical model. And polyurethane is chosen to be the combustion material in the study. To verify the work, temperature and species concentrations will be compared with the data calculated by direct numerical simulations (DNS) which is considered as "Numerical experiment". Next, N-Gas model is used to evaluate the toxicity of smoke and values of FED are calculated to investigate the influence of differential diffusion on smoke toxicity evaluation. Finally, some conclusions obtained from the study are presented.

2 Methodology

In this section, the practical CFD tool FDS was used. FDS is a free and open source code developed by the National Institute of Standards and Technology in the USA (NIST). It solves the Navier–Stokes equations for low Mach number flows with an emphasis on smoke and heat transport from fires [14]. The effective species Lewis numbers are obtained derived from the species and energy transport equations. And an effective Lewis number model which depends on turbulent Reynolds numbers is proposed.

3 Transport equation

The transport equation for species i can be written as

$$\frac{\partial}{\partial t}(\rho Y_i)+\nabla\cdot(\rho Y_i\mathbf{u})=\nabla\cdot(\rho D_i\nabla Y_i)+m_i \quad (1)$$

where ρ is the mixture density, $\boldsymbol{u}$ is the velocity vector, Y_i, D_i and m_i are the mass fraction, the mixture-averaged diffusivity and the production rate of species i, respectively.

Neglecting the temporal variations in pressure, the Low Mach number transport equation approximation for energy can be written as [14]

$$\frac{\partial}{\partial t}(\rho h_s)+\nabla\cdot(\rho h_s\mathbf{u})=\dot{q}'''+\nabla\cdot(\rho\alpha\nabla T)+h_f \ , \quad (2)$$

where h_s is the sensible enthalpy, T is the temperature, α is the mixture thermal diffusivity, h_f is the heat fluxes for diffusive and radiative, $\dot{q}'''$ is the heat release rate. And $\dot{q}'''$ can be expressed as

$$\dot{q}'''\equiv-\sum m_i\Delta H_{f,i} \ , \quad (3)$$

where $\Delta H_{f,i}$ is the heat of formation of species i.

Fig. 1 shows the contribution from each term in Eq. (2). Each contribution is normalized by the largest absolute value of all contributions. Obviously, the last term on the right side of Eq. (2) has negligible contribution in comparison to the other terms. The energy equation can be simplified as

$$\frac{\partial}{\partial t}(\rho h_s)+\nabla\cdot(\rho h_s\mathbf{u})=\nabla\cdot(\rho\alpha\nabla T)+\dot{q}''' \ . \quad (4)$$

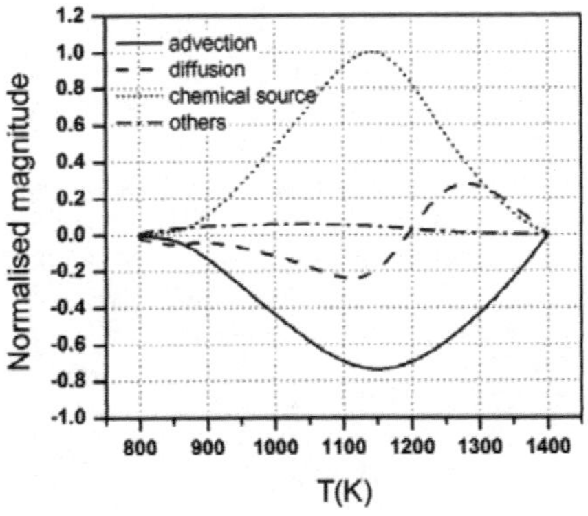

Figure 1. Contribution from each term in the energy equation as a function of temperature.

4 Effective lewis numbers

The use of the Reynolds-averaged form of the Navier-stokes equation (RANS) in CFD models allowed us to obtain the information about the ensemble average of the transported quantity. Hence, based on Eq. (1) and Eq. (4), the Reynolds-averaged species and energy equations can be written as (after applying the Favre filter)

$$\frac{\partial}{\partial t}(\bar{\rho}\tilde{Y}_i)+\nabla\cdot(\bar{\rho}\tilde{Y}_i\tilde{\mathbf{u}})=\nabla\cdot\left[\bar{\rho}(D_i+D_M)\nabla\tilde{Y}_i\right]+\bar{\dot{m}}_i \ , \quad (5)$$

$$\frac{\partial}{\partial t}(\bar{\rho}\tilde{h}_s)+\nabla\cdot(\bar{\rho}\tilde{h}_s\tilde{\mathbf{u}})=\nabla\cdot\left[\bar{\rho}(\alpha+D_T)\nabla\tilde{T}\right]+\bar{\dot{q}}''' \ , \quad (6)$$

where D_M and D_T denote the Reynolds-averaged and the Favre average, respectively. D_M and D_T are the eddy diffusivity for species and thermal. From Eq. (5) and Eq. (6), a formula for the effective species Lewis numbers can be obtained

$$Le_{i,eff}=\frac{D_{heat}}{D_{mass}}=\frac{\alpha+D_T}{D_i+D_M} \ , \quad (7)$$

where D_{mass} and D_{heat} represents the mass diffusivity and the thermal diffusivity, respectively.

In LES, the mass diffusivity and the thermal diffusivity are taken from the molecular and turbulent viscosities divided by the turbulent Schmidt number Sc_t and turbulent Prandtl number Pr_t, respectively. And Pr_t and Sc_t are assumed to be constant for a given scenario and the default value are 0.5 for both [14]. That means in LES,

$$Le_{i,eff}=\frac{(\mu+\mu_t)/\mathrm{Pr}_t}{(\mu+\mu_t)/Sc_t}=1 \ , \quad (8)$$

where μ and μt are the dynamic and eddy viscosity, namely LES adopts the unity Lewis number for each species actually.

In the present study, in order to calculate the unclosed terms in Eq. (7), the definition of the mixture-average binary Fickian diffusion [20] is used. The species mixture-averaged diffusivity can be written as

$$D_i = \frac{\sum_{j=1, j\neq i}^{N} X_j W_j}{\overline{W} \sum_{j=1, j\neq i}^{N} X_j / D_{ji}} , \tag{9}$$

where N is the total number of species, $\overline{W}$ is the mean molecular weight of the mixture, X_j and W_j are the mole fraction and the molecular weight of the jth species, respectively. D_{ji} is the binary diffusion coefficient of species j in species i and is defined as following:

$$D_{ji} = \frac{3}{16} \frac{\sqrt{2\pi k_B^3 T^3 / m_{ji}}}{P\pi\sigma_{ji}^2 \Omega^{(1,1)^*}} , \tag{10}$$

where k_B is the Boltzmann constant, m_{ji} is the reduced molecular mass for the (j,i) species pair, P is the pressure, σ_{ji} is the reduced collision diameter, and $\Omega^{(1,1)^*}$ is the collision integral. The detailed information of the binary diffusion coefficient can be found in the Ref. [21].

In Eq. (9), the round-off is accumulated in roughly the same way in both the numerator and denominator, and thus the formula is undefined if the mixture is exactly a pure species. So a small number ε is added to the actual mole fraction to overcome this difficulty. We assume in Eq. (9) that

$$X_j = \hat{X}_j + \varepsilon , \tag{11}$$

where ε equals to 10^{-12}.

As for the mixture thermal diffusivity, it can be computed from the thermal conductivity which is calculated by a combination averaging formula:

$$\alpha = \frac{\lambda}{\rho c_p} , \tag{12}$$

$$\lambda = \frac{1}{2}\left(\sum_{i=1}^{N} X_i \lambda_i + \frac{1}{\sum_{i=1}^{N} X_i / \lambda_i} \right) , \tag{13}$$

where λ_i is the thermal conductivity of the ith species. All the thermodynamic properties and the transport properties used to calculate the diffusivity can be obtained from the JANAF tables [22].

The method mentioned above which is used to solve molecular diffusion problems has been successfully applied in different researches such as Ref. [20].

Using the Deardorff's model [14] to calculate the remaining two unclosed terms in Eq. (7), the eddy diffusivity can be written as

$$D_T = \frac{\mu_t}{\Pr_t} = \frac{C_v \Delta \sqrt{k_{sgs}}}{\Pr_t} \quad D_M = \frac{\mu_t}{Sc_t} = \frac{C_v \Delta \sqrt{k_{sgs}}}{Sc_t} , \tag{14}$$

where Δ is the integral length scale, k_{sgs} is the kinetic energy, and C_υ is the model constant which is set to the literature value 0.1.

5 Proposed model

From the previously literature, it is well known that diffusion of temperature and species is dominated by turbulence level [23]. Hence, a choice of the effective Lewis numbers depend on the turbulent Reynolds number is made.

From Eq. (7), the effective Lewis number can be obtained for each species

$$Le_{i,eff} = \frac{1 + D_T/\alpha}{D_i/\alpha + D_M/\alpha} = \frac{1 + D_T/\alpha}{1/Le_i + D_M/\alpha} , \tag{15}$$

where Le_i is the species molecular Lewis number.

Using the definition of the mixture Prandtl number Pr and the turbulent Reynolds number Re_T, the ratio of the eddy to the thermal diffusivities and to the mass diffusivities can be written as

$$\frac{D_T}{\alpha} = \frac{1}{10}\frac{\Pr}{\Pr_t}\mathrm{Re}_T \quad \frac{D_M}{\alpha} = \frac{1}{10}\frac{\Pr}{Sc_t}\mathrm{Re}_T , \tag{16}$$

where $\mathrm{Re}_T = \mu' l / \upsilon$, μ' is the root-mean-square velocity fluctuation which can be written as $\mu' = k^{1/2}$, k is the kinetic energy, l is the characteristic length, and υ is the kinematic viscosity [24].

So, the species effective Lewis number becomes

$$Le_{i,eff} = \frac{1 + C_1 \mathrm{Re}_T}{1/Le_i + C_2 \mathrm{Re}_T} , \tag{17}$$

where

$$C_1 = \frac{1}{10}\frac{\Pr}{\Pr_t} \quad C_2 = \frac{1}{10}\frac{\Pr}{Sc_t} . \tag{18}$$

It is difficult to evaluate the parameters C_1 and C_2 exactly, but an estimate of them can be provided. In FDS, the default value of turbulent Prandtl number and turbulent Schmidt number are 0.5 for both and Pr is set to 0.7 [14]. From the discussion above, the two parameters have the same value 0.14 approximately. Hence, the species effective Lewis number can be written as

$$Le_{i,eff} = \frac{1 + C_T \mathrm{Re}_T}{1/Le_i + C_T \mathrm{Re}_T} , \quad C_T \approx 0.14 . \tag{19}$$

In Eq. (19), Le_i and Re_T are the inputs. Le_i can be obtained from literature survey, Re_T is a known quantity from the simulations and C_T here is used as a fitting coefficient and should be modified in specific cases.

6 Numerical set-up

A 1/5 scale vertical shaft model which was first chosen to research about the smoke in vertical shaft by Marshall [25] was considered. In the scaling model, there are five floors, each floor is 0.66 meters. The doors on the ground floor and the top floor are open. The height of the door is 0.3 m and 0.44 m, respectively.

A geometrically simplified model (2D) is illustrated in Fig. 2. The polyurethane foam (polyurethane is defined by the chemical formula $C_{25}H_{42}O_6N_2$ in this study [26]) is chosen as the material and is set in the corner of the fire compartment. And there are 25 thermocouples in the shaft, they are set every 0.13 m and the first one is set at 0.08 m.

To demonstrate that the simulations are sufficiently well-resolved, mesh sensitivity study is presented for all simulations. In the mesh resolution study, a relatively coarse mesh is used first, and then gradually refines the mesh until there are no appreciable differences in our results. In the case of large eddy simulations, three different grids are used, grid 1 (1 cm), grid 2 (2.5 cm) and grid 3 (5 cm).

Fig. 3 shows the average temperature in the vertical shaft simulated with LES. Grid 1 and grid 2 depart immediately from grid 3, indicating that 5cm is insufficient in this case, but are in close agreement with each other. The turbulent nature of the flow makes it inherently unstable, and there are no appreciable differences between grid 1 and grid 2 demonstrating that the simulation is being well-resolved with grid 2. And 5 mm was selected for DNS simulation after doing the same mesh sensitivity study.

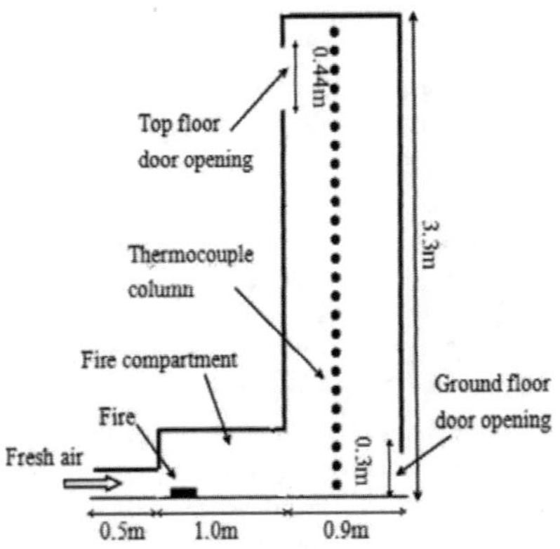

Figure 2. Vertical shaft model.

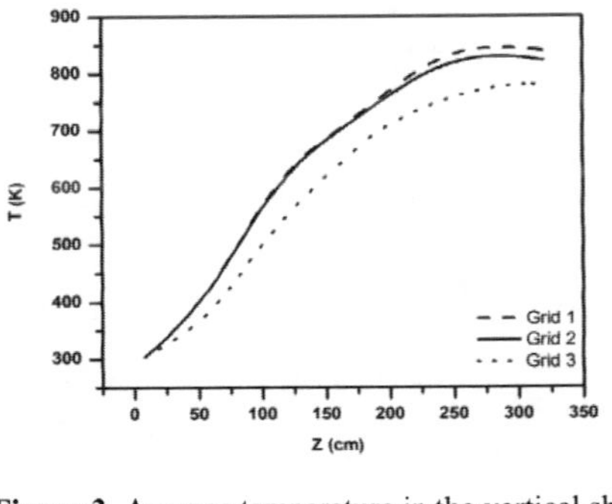

Figure 3. Average temperature in the vertical shaft.

7 Results and discussion

In this section, the fire described above is simulated with the effective and the unity species Lewis numbers. The temperature and the main species mass fractions are compared with the DNS data in validation.

8 Mean results

In order to observe the comparison results, mean results for temperature and species mass fractions on the center line of the shaft are presented in Fig. 4. The DNS data are indicated with symbols while the results with effective Lewis numbers or unity Lewis numbers are presented with solid and dashed lines, respectively.

Obviously, the temperature profile is well captured by the simulations with the effective Lewis numbers and the same applies for the species mass fractions profiles. In Fig. 4, it is observed that the temperatures and mass fractions of CO_2, CO and H_2O are overestimated while the mass fractions of $C_{25}H_{42}O_6N_2$ and O_2 are underestimated when differential diffusion effects are not taken into account. Moreover, the peak temperature simulated with the effective Lewis numbers is at the position of about z = 268 cm, coinciding with the maximum values for CO_2, CO and H_2O mass fractions. On the other hand, if differential diffusion effects are neglected, the temperature distribution is not well captured which directly leads to the discrepancies in the species mass fractions.

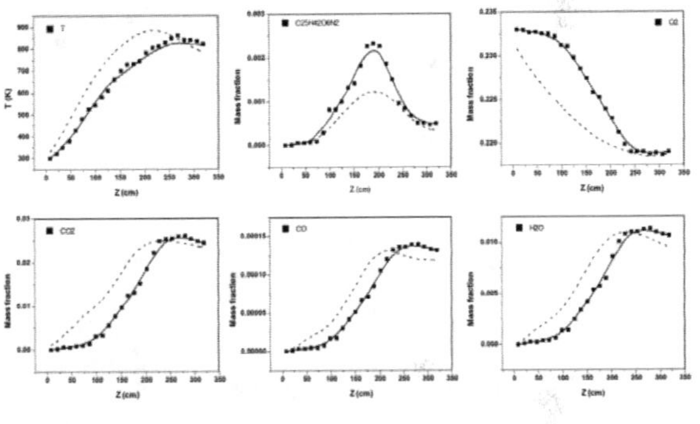

Figure 4. Temperature and species mass fractions at the center line of shaft. DNS data: symbols; effective Lewis numbers: solid lines; unity Lewis numbers: dashed lines.

9 Conditional mean results

From the discussion above, species concentration is influenced by temperature, especially for CO_2, CO and H_2O. In reference [27], the conditional means of species concentrations with regard to temperature $\langle Y_i | T \rangle$ for average of relatively steady flame are calculated. Fig. 5 shows the conditional mean mass fractions of CO_2, CO and H_2O as a function of temperature at the location of z = 268 cm where temperature reached the peak value. The DNS data are indicated with solid lines while the results with effective Lewis numbers or unity Lewis numbers are presented with dashed and dotted lines, respectively.

As mentioned earlier, the simulation results can be more accurately captured by LES method with an appropriate change in Lewis numbers of different species. Compare the data of these three groups, it is observed that the concentration of combustion products will reach a higher value at the same temperature when simulated with the unity Lewis number. In the case of differential diffusion effects are taken into account, there are also exist some differences when compared results with DNS data. To claim the errors, the L2-norm errors of the species conditional mean mass fractions between DNS and LES are calculated. The error can be written as (take carbon dioxide as an example):

$$L2=\sqrt{\frac{1}{N}\sum_{i=1}^{N}\left(Y_{CO_2}^{LES}(\mathbf{Le}^*,T_i)-Y_{CO_2}^{DNS}(T_i)\right)^2}, \tag{20}$$

where $Y_{CO_2}^{DNS}(T_i)$ corresponds to the interpolated value of $\langle Y_{CO_2}|T\rangle$ from DNS at $T = T_i$, and $Y_{CO_2}^{LES}(\mathbf{Le}^*,T_i)$ corresponds to the value of $\langle Y_{CO_2}|T\rangle$ from LES at $T = T_i$ with the effective Lewis numbers. Fig. 6 shows the L2-norm errors between the DNS and LES (with unity or effective Lewis numbers). Obviously, the errors between DNS and LES with effective Lewis numbers are much smaller than that with unity Lewis numbers for species in each case. And all the errors are less than 10% which is acceptable to engineering application [28]. When unity Lewis number is used, the errors of the reactants are bigger than 10%, especially for oxygen.

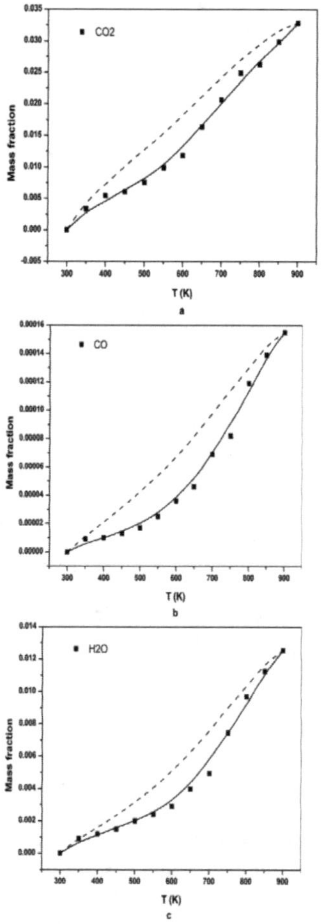

Figure 5. Conditional mean species mass fraction profiles as a function of temperature. DNS data: symbols; effective Lewis numbers: solid lines; unity Lewis numbers: dashed lines.

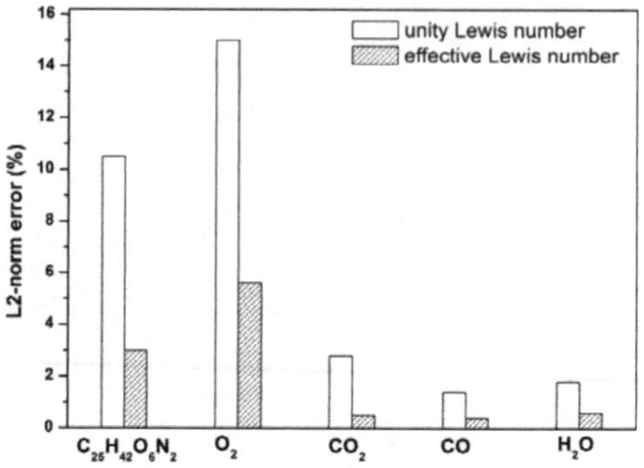

Figure 6. L2-norm errors for species between DNS and LES with the effective or unity Lewis numbers.

10 Smoke toxicity evaluation

In this sub-section, N-Gas model was used to evaluate the toxicity of the smoke. The 3-gas value can be written as [29]:

$$FED=\frac{m[CO]}{[CO_2]-b}+\frac{21-[O_2]}{21-LC_{50}(O_2)}, \tag{21}$$

where $[X_i]$ represents the concentration of species X_i (ppm); $LC_{50}(O_2)$ represents the lethal concentration of oxygen (%), and it is taken as 5.4% here; based on the empirical values, m = -18, b = 122000. From the comparison of the results of animal experiments with the N-Gas model it is shown that 50% animals will die when the value of FED is equal to 1.

Fig. 7 shows the FED values at the center line of shaft. Apparently, the values of FED are different with the effective or unity Lewis numbers. When differential diffusion effects are not considered, values of FED are overestimated and the largest differences are observed at $z \geq 200$ cm where fire smoke toxicity has reached the critical value of death. On the other hand, the values of FED are always less than 1 by the simulation with the effective Lewis numbers.

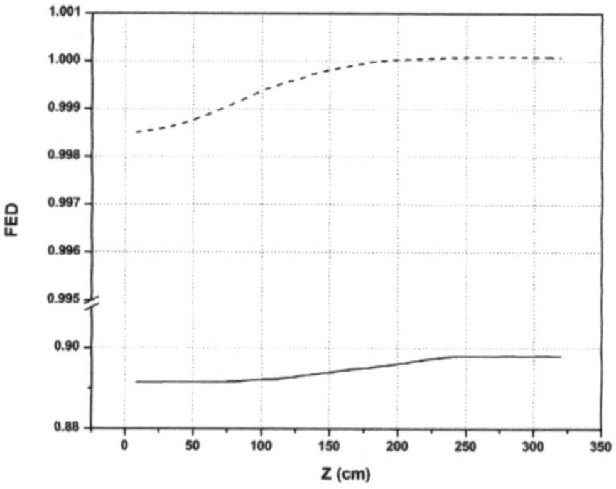

Figure 7. FED values at the center line of shaft. Effective Lewis numbers: solid lines; unity Lewis numbers: dashed lines.

11 Conclusions

With the emergence of new building materials and the use of existing materials, smoke toxicity evaluation becomes one of the most important topics in related fields. There is an increasing tendency to use numerical simulation methods to predict the composition and concentration of smoke which are needed for toxicity evaluation model. Large eddy simulation is considered to be one of the most potential methods to simulate building fires. Extending the ability of LES method to predict the smoke concentration accurately becomes very meaningful.

In the present study, the effective Lewis numbers of species derived from the species and energy transport equations have been presented to incorporate differential diffusion in CFD simulations. The study was verified by simulating a polyurethane fire in a 1/5 scale vertical shaft and the numerical results were compared with the data derived from DNS. Temperature and mass fractions of CO_2, CO and H_2O are overestimated while the mass fractions of $C_{25}H_{42}O_6N_2$ and O_2 are underestimated when simulated with unity Lewis number. On the other hand, temperature and species mass fractions are in well agreement with DNS data when differential diffusion effects are taken into account. Values of FED are calculated based on the data simulated with the effective or unity Lewis numbers using N-Gas model. Results indicate that the prediction model will overestimate smoke toxicity when differential diffusion effects are neglected. Moreover, a model derived from the transport equations for the effective Lewis number which depends on turbulent Reynolds numbers is proposed. The model is found to be applicable in turbulent flames. More data are needed to fully evaluate the model, for example, flames with Lewis number of fuel less than unity shall to be compared to the model. Future work will be concentrated on the fully validation and the applications of the model.

This article provides a method for CFD modeling to capture the behavior of smoke in building fires more accurately with an appropriate change in the Lewis number of species. The modified CFD model can be used to predict the combustion products of materials in building fire and provide more accurate data for the gas toxicity evaluation.

Acknowledgment

This work was sponsored by the National Natural Science Foundation of China (51176181), the National Basic Research Program of China (2012CB719704), and the Research Fund for the Doctoral Program of Higher Education (20123402110047, 20133402110010), for which the authors would like to express their gratitude.

References

1. S. Nazare, B.K. Kandola, A.R. Horrocks, Smoke, CO and CO2 measurements and evaluation using different fire testing techniques for flame retardant unsaturated polyester resin formulations, J Fire Sci., **26**, 215-242 (2008)
2. J.H. Sun, L.H. Hu, Y. Zhang, A review on research of fire dynamics in high-rise buildings, Theoretical and Applied Mechanics Letters., **3**, 1-13 (2013)
3. W.C. Fan, J.H. Sun, S.X. Lu, Fire risk assessment methodology, 1 st ed. Beijing, China: Science press, **2**, (2004)
4. G.E. Hartzell, Overview of combustion toxicology, Toxicology, **115**, 7-23 (1996)
5. Y. Alarie, Toxicity of fire smoke, Crit Rev Toxicol, **32**, 259-289 (2002)
6. J. J. Xing, Y.Jiang, L.W. Pan, An approach for predicting the toxicity of smoke, Journal of Safety Science and Technology, **8**, 72-87 (2013)
7. P. Blomqvist, T. Hertzberg, H. Tuovinen, K. Arrhenius, L. Rosell, Detailed determination of smoke gas contents using a small-scale controlled equivalence ratio tube furnace method, Fire Mater., **31**, 495-521 (2007).
8. A. A. Stec, T. R. Hull, K. Lebek, J. A. Purser, D. A. Purser, The effect of temperature and ventilation condition on the toxic product yields from burning polymers, Fire Mater., **32**, 49-60 (2008)
9. J.L. Sebaugh, Guidelines for accurate EC50/IC50 estimation, Pharm Stat., **10**, 128-134 (2011)
10. G.L. Neson, G.E. Hartzell, B.C. Levin, Group discussion: New research avenues in combustion toxicology, Toxicology, **115**, 185-200 (1996)
11. B.C. Levin, New research avenues in toxicology: 7-gas N-gas model, toxicant suppressants, and genetic toxicology, Toxicology, **115**, 89-106 (1996)
12. A.A. Stec, T.R. Hull, Assessment of the fire toxicity of building insulation materials, Energ Buildings., **43**, 498-506 (2011)
13. K.T. Paul, T.R. Hull, K. Lebek, A.A. Stec, Fire smoke toxicity: The effect of nitrogen oxides, Fire Safety J., **43**, 243-251 (2008)
14. K. McGrattan, S. Hostikka, R. McDermott, Fire Dynamics Simulator, Technical Reference Guide, (2014)
15. J.D. Regele, E. Knudsen, H. Pitsch, G. Blanquart, A two-equation model for non-unity Lewis number differential diffusion in lean premixed laminar flames, Combust Flame., **160**, 240-250 (2013)
16. G. Maragkos, P. Rauwoens, B. Merci, A new methodology to incorporate differential diffusion in CFD simulations of reactive flows, Combust Flame., **160**, 1903-1905 (2013)
17. G. Maragkos, P. Rauwoens, B. Merci, Differential diffusion effects in numerical simulations of laminar, axi-symmetric H2/N2–air diffusion flames, Int J Hydrogen Energ., **39**, 13285-13291 (2014)
18. L. Smith, R. Dibble, L. Talbot, R. Barlow, C. Carter, Laser Raman scattering measurements of differential molecular diffusion in turbulent nonpremixed jet flames of H2/CO2 fuel, Combust Flame., **100**, 153-160 (1995)
19. N.R. Marshall, The behaviour of hot gases flowing within a staircase, Fire Safety J., **9**, 245-255 (1985)
20. R.J. Kee, F.M. Rupley, J.A. Miller, CHEMKIN Release 4.0, Reaction Design, (2004)

21. J.O. Hirschfelder, Ch.F. Curtiss, R.B. Bird, Molecular Theory of Gases and Liquids, New York: Wiley, (1954)
22. M.W. Chase jr., NIST-JANAF Thermochemical Tables, (1998)
23. A.J. Aspden, M.S. Day, J.B. Bell, Turbulence–flame interactions in lean premixed hydrogen: transition to the distributed burning regime, J. Fluid Mech., **680**, 287-320, (2011)
24. Zhang Zhaoshun, Zui Guixiang, Xu Chunxiao,Theory and Modeling of Turbulence. Tsinghua University Press, Beijing, China, (2005)
25. N.R. Marshall, Air entrainment into smoke and hot gases in open shafts, Fire Safety J., **10**, 37-46 (1986)
26. J. Floyd, G. Forney, S. Hostikka, T. Korhonen, R. McDermott, K. McGrattan, Fire Dynamics Simulator, User's Guide, (2012)
27. B. Savard, G. Blanquart, An a priori model for the effective species Lewis numbers in premixed turbulent flames, Combust Flame., **161**, 1547-1557 (2014)
28. Chen Jianguo, Chen Haixin, Fu Song, Two typical phenomena from the numerical simulation of fire and smoke transport in a gymnasium fire, Chinese Sci Bull., **51**, 1521-1525 (2006)
29. B.C. Levin, New approaches to toxicity: A seven-gas predictive model and toxicant suppressants, Drug Chem Toxicol., **20**, 271-280 (1997)

MATEC Web of Conferences **64**, 07006 (2016) DOI: 10.1051/matecconf/20166407006
ICCMP 2016

Green Sharing: The Proposed Criteria in Green Building Standards to Promote the Usage of Natural Handicrafts in Building Materials

Aracha Krasae-In

Kasetsart University, Thailand

Abstract. Sustainable development has been a great challenge to the building and construction industry for decades. There have been many initiatives and attempts to create sustainability for the industry through the concept of the Green Building certificate in order to reduce the impact to environment and society while promoting better living conditions of the people involved in the project. This paper aims to examine all three aspects of sustainability; economy, environment and society, in the building and construction industry by proposing new criteria for the green building certificate. This will create opportunities for the community based handicraft building products to be specified and purchased to be used in the modern building and construction industry and share the economic value to the community.

1 Introduction

It is already known that the building and construction industry is one of the major causes of pollution and an impact to the environment. According to the EPA, the industry has created approximately one third of annual greenhouse gases and also consumed a large amount of material as well as produced an enormous amount of waste along its life time. This includes the extraction of construction materials, the transportation and production of material, the construction process and the operation of the building.

However, the scope of the green building sustainability was limited to the direct benefit of the construction and building. According to the research by McGraw-Hill [1], the most important reasons for the green building project in terms of social factors were "to promote greater health and well-being" and "to encourage sustainable business practices" while the most important in terms of environmental reasons were "to lower energy consumption" and "reduce greenhouse gas emissions".

The practice of sustainability was far beyond the minimalizing of impact but also to promote the betterment of society and environment. On the other hand, the businesses should share their profits to promote genuine sustainability.

2 Sustainability and green building

Sustainability is currently one of the world's most focal issues. Pressure from environmental concern from the changing of global climate, natural pollution and scarcer resources, the concept of sustainability is not only about good will, but a necessity for everyone.

According to the United Nation's World Commission on Environment and Development or the Brundtland's report [2], the word "sustainability" was defined as the goal for sustainable development which is identified as "The Three Pillars of Sustainability" including economic development, social development and environmental protection.

As mentioned earlier, the building and construction industry, which has been largely responsible for the impact to society and environment, had adopted the idea of sustainability. Initiatives were implemented in many countries in order to reduce the impact from design and construction which have developed into the green building standard and rating system all over the world.

Efforts towards sustainable building have been accompanied by the development of methods, tools, and systems to assess the sustainable performance of buildings. Kibert [3] defined the "High performance green building" as facilities designed, built, operated, renovated, and disposed of using ecological principles for the purpose of promoting occupant health and resource efficiency plus minimizing the impacts of the built environment on the natural environment.

According to OECD [4], generally, sustainability of the building or green building covered many issues from the design of the building, site selection, the specification of the materials and the efficiency of building system; to the construction which starts from the site preparation, the construction protection, and the minimization of the construction waste. The green building also has included the operation and resources used after the building was occupied which consisted over its operation life-time.

When comparing the globally well-known Green Building standard or rating system, it is important to consider the criteria within the Green Building rating system. For example, LEED V 4.0 [5] which is currently an active version, is comprised of 6 categories including; the sites and sustainability, the location and transportation, the water efficiency, the indoor environment quality, the material and researches, and the energy and atmosphere. These criteria for each program have its coverage in all aspects within the design, construction and operation of the building. Beside LEED's criteria, the other leading green building assessment tools including BRE Environmental Assessment Method (BREEAM, United Kingdom), Green Building Council of Australia Green Star (GBCA, Australia), or DGNB (Germany), might have a degree of dissimilarity. However, they were created on the similar basis of sustainability as shown in the image below.

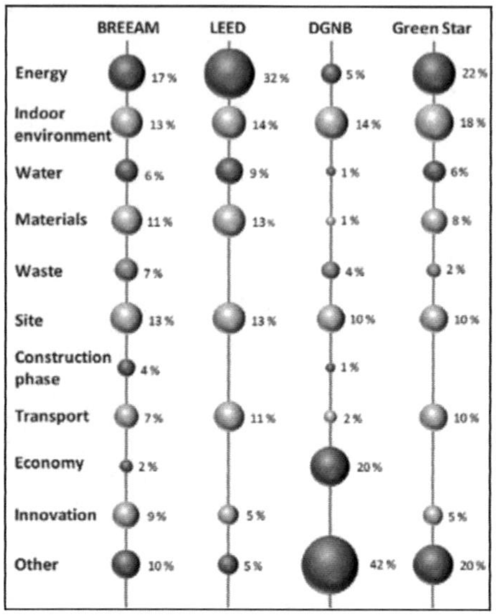

Figure 1. Comparison of green building standard [6].

Jian Zuoa and Zhen-Yu Zhao [7] reported that the focused aspect of the Green Building assessment tools was mainly placed on environmental aspects of sustainability shown in the criteria including energy efficiency, water efficiency, resource efficiency and greenhouse gas emission reduction.

However, the improvement of these assessment tools in the last decade showed the growing concerns of social sustainability in buildings. In the construction context, social sustainability mainly covers the better quality of living, occupational health and safety, and the future professional development opportunities, while in building context, social sustainability means providing a healthy and safe environment to all stakeholders including building occupants, residents, the construction personnel and other users and operators of the property which should be taken into account during the sustainable design process [7]. Moreover, the corporate social responsibility or CSR of the corporations was also the accelerating factor which the construction or real estate development companies often featured in their sustainability report [8].

Valdes-Vasquez and Klotz [9] suggested that social sustainability should be taken into consideration in construction projects right from the planning stage, including: engaging stake holders including end users, assessment of social impacts, and consideration of the local community.

Although the building design and construction of high performance buildings was dynamic and evolving, economic development aspects of sustainability in the green building were still left over. In LEED V4.0, the criteria which had an intention that involved the development of local community or the sharing to domestic economy were minimal. For example, the "Regional Material" in LEED V 4.0which provided the project an extra factor to calculate the credit in Material and Resources Categories in the case of specified and purchased building products sourced (extracted, manufactured, purchased) within 100 miles. This condition aimed to reduce the energy and carbon-emissions during transportation rather than promoting the surrounding economy.

In order to fulfil the pillar of sustainability in all aspects, the green building standard and rating system should considered an option for the project to share an economic value along with the existing guidelines which aimed to both reduce as well as promote an environmental and social quality. One option is to elevate the regional material concept to meet the sustainability needs of the society, especially in Thailand where one particular concern is the sustainability of the rural community. There is an option for the green building to share its value to others.

3 Natural handicrafts for architecture

3.1 Natural materials

The definition of natural materials is simple; anything that is available in the nature could be identified as natural materials. However, in the domain of sustainability, 'natural materials' means an organic natural resource which could be renewable through naturally farming or plantation to overcome usage and consumption [10]. Rapidly renewable material is defined by the USGBC as a material that was able to regenerate itself in 10 years or less [5].

According to LEED V 4.0, the specification of natural material has been placed in the category of Material and Resources (MR) with the credit of Building product disclosure and optimization - sourcing of raw materials. The natural material was identified as (1) Bio-based materials which must meet the Sustainable Agriculture Network's Sustainable Agriculture Standard, their raw materials must be tested using ASTM Test Method D6866 and be legally harvested, as defined by the exporting and receiving country, and(2) Wood products which must be certified by the Forest Stewardship Council (FSC) or a USGBC-approved equivalent.

However, there were numerous types of natural materials that could be identified as green materials since they could be renewed rapidly and legally harvested.

There were also the traditional buildings that were made of the natural materials which provided many benefits to sustainability of the construction. Stefan Dirlich [11] mentioned that the advantages included the utilisation of natural building materials, relatively low energy content of building material, excellent energy-efficiency during the phase of use and natural thermal insulation.

The study by Panyakaew and Fotios [12] showed the performance of insulation materials in the property. Bagasse, rice hulls and coconut coir were shown to offer the greatest potential for manufacturing into thermal insulation products.

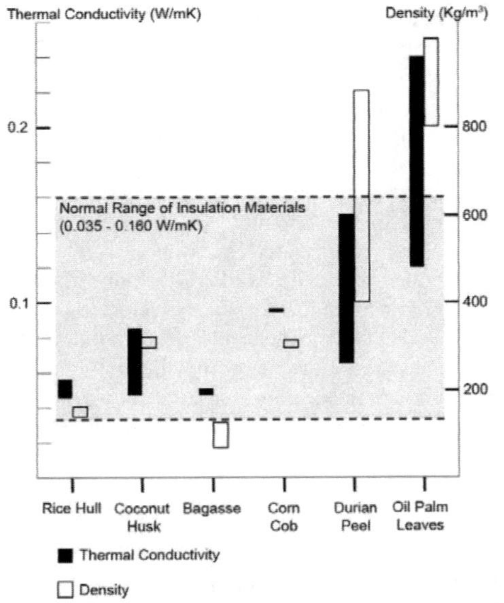

Figure 2. Comparison of thermal conductivities and densities of insulation board made from agricultural waste materials [12].

As mentioned earlier in this paper; most of the green building assessment tools had mainly focused on minimizing the environmental impact. Subsequently they missed an opportunity to promote sustainable business by promoting the usage of natural material.

3.2 Domestic handicrafts

As mentioned above, traditional buildings were made of natural materials such as; bamboo, rattan and vetiver, which can be found in the region, were produced with handcrafting. The natural building materials that were used in traditional construction were mostly sourced from the surrounding area or near the construction sites. The regional building materials also had relatively high shares of handwork positively affecting the local labour rmarket. These were generally user-friendly and easy-to-use materials and avoided damage to historic buildings by using existing technologies and materials. These materials adapted to most adverse conditions and enhanced the value of the building [11].

In Thailand and as in many Asian countries, handicrafts were not only the production or business, but have been the bond of the communities, creating jobs and generating substantial income in many rural areas of the country. Craftsmanship, for handicraft products, was a development in human society that transformed material into form and function by a community originating from the same location as the material. It contains knowledge and skills of the producers in the transforming of natural materials locally found such as bamboo, sage, rattan or even clay and metal into products [13].

However, after industrialisation, the handicraft was overcome by the mass production system. In order to compete with industrialised products, the craft producers had to sell their handmade products at a lower price [14]. The shift to mechanisation also changed people's lifestyle and their needs for products, therefore handicraft products gradually lost consumers. Unfortunately, handicraft producers were not aware of such avolatile threat and were unable to adapt themselves to meet the new demands from modern consumers.

3.3 Case of thatch

Thatch is a conventional vernacular construction material in many tropical regions. In Thailand, thatch was made by the wickerwork of Nipa Palm Leaves; the material that is made of natural leaves that grow along the mangroves and delta. It was once one of the major materials used in Thai traditional houses as a roofing material.

The decline of thatch demand consequentially led to the abandon of folk wisdom in the production of thatch as well as deforestation of the delta area since villagers did not perceive the Nipa Palm as their major source of income. Besides that, the income from the making of Nipa Palm thatch is relatively lower than other kinds of work especially when compared to the permanent workers in the factory sector or other service industries. This automatically triggered the collapse of the communities as a result of the abandonment of the villages by the younger generations.

However, thatch roofing has shown its ability tobe an excellent heat guard. From the experiment, the thatch was able to reduce 4.3C of the temperature in the under shade, compared to the normal galvanized steel roof. This is a substantial figure in terms of heat reduction, furthermore it consequentially reduced the energy used to cool the house by air-condition [15].

It was proven that the performance of the material (thatch) itself was able to help the design to achieve the required energy performance and the green certificates. But this data was unrecognized by the design and construction industry. Moreover, the industry would prefer to purchase an industrial product due to their convenience and ease in acquiring information about the products. As a result, there would be no business for natural handicraft building materials or products even if they possessed the necessary quality for green building.

In order to help the production of handicrafts and the community, there was a proposed promotional program that endorsed the design and construction industry to specify and purchase these products. One of the tools that could help promoting them was the green building mechanisms that have the criteria required for the building to specify and utilize handicraft products made from renewable natural material.

4 Green sharing as a standard

In order to share an economic value to the society while promoting the environmentally friendly and sustainable business, the authorities of the green building certificates and assessment tools should create criteria requiring that any project pursuing the certification must specify and purchase a certain amount of the natural handicraft building materials.

In Thailand, the components for this initiative already exist. On one side, TGBI or Thai Green Building Institute has created Thailand's green building assessment tool called "TREES Rating System" [16] which shares similarities with the LEED 2009 rating system. In this rating system, there is one criterion that demands the project to specify and purchase domestic products with an intention to promote the consumption of domestic building materials and reduce the energy used by transportation. This new credit that aimed to promote the natural handicraft building materials could be derived from the same method.

On the other hand, there are numerous handicraft producers that make the products for architectural projects. There was also a government initiative that intended to identify, develop and promote the handicraft producers by creating the standard for community products or Thai Community Product Standard (TCPS) which covered any type of products from food to furniture that could prove their authenticity as a community and therefore their consistency of production.

Figure 3. Mark of the standard of community product by Thailand Industrial Standard institute.

As of October 2015, there were 1,360 categories of standard that registered, but none of them have registered under building material or construction materials. These two components could create the mechanism that allows the community businesses to gain the economic value of the building and construction industry and thusly enable it to complete the basic pillar of sustainability.

In international cases, this initiative could be implemented through the green building rating system too. However, the verification mechanism to identify through natural community building products has to change. There were several international organizations, either governmental or non-profit private, that provided the verifying tools; for example, Fair Trade Organization, World Craft Organization. Green building organizations should work with assistance from those organizations in order to create the best mechanism for the green building to specify and purchase the products from the appropriate producers or suppliers.

5 Conclusion

The sustainability of the building and construction industry could be created not only from minimizing or eliminating the impact from the activities of the business, but also from sharing the economic value to the environment and society by specifying and selecting the handicraft products made from natural renewable material from the right producers.

To endorse the sharing, green building institutes should consider the idea of using the natural material handicraft as their credit in the material section of the green building standard. The credit award procedure should be studied in order to assure the integrity and transparency of the credit as well as the guidelines for the project developer, architect and green building practitioner to proceed to the credit.

On the other hand, the handicraft producers should be informed about the need for their product for the green building so that they can develop the products that are able to meet the requirements of the building and the criteria of the green building standard.

References

1. McGraw-Hill Construction, SmartMarket Report, (2013)
2. G. Brundtland, M. Khalid, S. Agnelli, S. Al-Athel, B. Chidzero, L. Fadika, V. Hauff, I. Lang, Ma Shijun, M.M. de Botero, M. Singh, S. Okita,Our Common Future (\'Brundtland report\'), (1987)
3. C.J. Kilbert, Green Buildings: An Overview Of Progress, Journal of Land Use, **19**, 2 (2004)
4. OECD, Design Of Sustainable Building Policies: Scope For Improvement And Barriers, (2002)
5. USGBC,LEED v4, LEED Reference Guide for Building Design and Construction, (2014)
6. P. Vladykova, Presentation in Chalmers Energy Conference 2013, (2013)
7. J. Zuoa, Z.Y. Zuoa, Green building research–current status and future agenda: A review, Renewable and Sustainable Energy Reviews, **30**, 271–281, (2014)
8. K.D. Wong, Q. Fan, Building information modelling (BIM) for sustainable building design Facilities, **31**, (3/4) (2013)
9. R. Valdes-Vasquez, L.E. Klotz, Social sustainability considerations during planning and design: A framework of processes for construction projects,J. Construct. Eng. Manage, **139**, 1 (2013)
10. J.G. Stead, W.E. Stead, Management for a Small Planet, (2009)
11. S. Dirlich, The Building Stock and Traditional Building Principles: Sustainability Assessment for Historic Buildings,Proceedings of the 1st International Conference on Building Sustainability

DOI: 10.1051/matecconf/20166407006

Assessment. Barcelos (Portugal): Green Lines Institute for Sustainable Development, (2012)
12. S. Panyakaew, S. Fotio, 321: Agricultural Waste Materials as Thermal Insulation forDwellings in Thailand: Preliminary Results, (2008)
13. V. Leesuwan, Thai Traditional Crafts, (1984)
14. V. Leesuwan, Wickerwork in Thailand, (1999)
15. Silpakorn University Research and Development Institute, Material and construction system in Thai vernacular house for sustainable living, (2005)
16. TGBI, TREES Rating System, (2009)

AUTHOR INDEX

Ahmad, S. K. 6
Ai, Sufen 40
Aji, Demas 36
Alam, M. S. 1
Benjaratthapong, Kannika 59
Bugala, Artur 49
Buyukyildiz, Muhammet 12
Cieslik, Marta 49
Czekala, Wojciech 49
Demir, Goksel 12
Dixit, Abhilasha 1
Du, Yu 45
Geng, Yunguang 26
Hadi, M. N. A. 54
He, L. 70
Imam, Z. S. 64
Janczak, Damian 49
Jiang, Yong 74
Jongtrakulsiri, Meatra 59
Khattab, Rania 64
Kozlowski, Kamil 49
Krasae-In, Aracha 81
Laksono, Endang W. 36
Li, Jing 30
Li, Xu 74
Lin, Xiong-Jie 30
Lin, Zhan-Xi 30
Liu, Bin 40
Liu, Lu 20
Marfuatun 36
Mishra, P. K. 1
Mohamed, O. A. 64
Ni, Yanhui 20
Oruc, Ilker 12
Ozcan, H. K. 12
Ozdemir, Huseyin 12
Pan, Longwei 74
Pektas, A. O. 12
Shao, En-Si 30
Shen, Yong 40
Shuling, Tang 26
Sun, Xixiang 45
Tang, Dazhen 26
Tang, Lihua 20
Usman, Halima 6
Uwanta, I. J. 6
Wang, Lei 74
Wangrakdiskul, Ubolrat 59
Wu, Yongqiang 20
Xu, Hao 26
Yee, H. M. 54
Yin, Lulu 74
Zbytek, Zbyszek 49
Zhao, Y. 70
Zhao, Yingmin 40
Zhou, Qing 40
Zhu, Zibin 20

EDP Sciences – Web of Conferences
17, Avenue du Hoggar
Parc d'Activité de Courtabouf
BP 112
F-91944 Les Ulis Cedex A
France

ISBN 978-1-5108-2269-6

CURRAN ASSOCIATES INC.
proceedings.com

9th EPS-QEOD Europhoton Conference on Solid-State, Fibre, and Waveguide Coherent Light Sources (EUROPHOTON 2020)

EPJ Web of Conferences Volume 243 (2020)

Online
30 August - 4 September 2020

edp sciences